The Principles of Skeletal Muscles

A Reference for Students of Physical Therapy, Medicine, Sports, and Bodywork

Chris Jarmey

HUMAN
KINETICS

First published in 2024 by
Lotus Publishing
Apple Tree Cottage, Inlands Road, Nutbourne, Chichester, PO18 8RJ, and
Human Kinetics
1607 N. Market Street, Champaign, Illinois 61820

United States and International
Website: **US.HumanKinetics.com**
Email: info@hkusa.com
Phone: 1-800-747-4457

Canada
Website: **Canada.HumanKinetics.com**
Email: info@hkcanada.com

Anatomical Drawings Amanda Williams
Text Design Medlar Publishing Solutions Pvt Ltd., India
Cover Design Chris Fulcher
Printed and Bound Replika Press Pvt Ltd., India

Acknowledgments
The publisher would like to thank the following people who have helped with the material in this book in many differing and invaluable ways. In no particular order, Dr. Daniel Quemby, Dr. Robert Whitaker, Emily Evans, and Cecilia Brassett.

Medical Disclaimer
This publication is written and published to provide accurate and authoritative information relevant to the subject matter presented. It is published and sold with the understanding that the author and publishers are not engaged in rendering legal, medical, or other professional services by reason of their authorship or publication of this work. If medical or other expert assistance is required, the services of a competent professional person should be sought.

British Library Cataloging-in-Publication Data
A CIP record for this book is available from the British Library

Library of Congress Cataloging-in-Publication Data
Names: Jarmey, Chris, author.
Title: The pocket atlas of skeletal muscles / Chris Jarmey.
Description: Chichester, England : Lotus Publishing ; Champaign, Illinois :
    Human Kinetics, 2024. | Includes index.
Identifiers: LCCN 2023015460 (print) | LCCN 2023015461 (ebook) |
    ISBN 9781718226951 (paperback) | ISBN 9781718226968 (epub) |
    ISBN 9781718226975 (pdf)
Subjects: MESH: Muscle, Skeletal--anatomy & histology | Atlas | Handbook
Classification: LCC QM100 (print) | LCC QM100 (ebook) | NLM WE 17 |
    DDC 612.7022/2--dc23/eng/20230523
LC record available at https://lccn.loc.gov/2023015460
LC ebook record available at https://lccn.loc.gov/2023015461

ISBN: 978-1-7182-2695-1
10 9 8 7 6 5 4 3 2 1

Contents

About this Book

This book is designed in quick-reference format to offer useful information about the main skeletal muscles that are central to sport, dance, exercise science, and bodywork therapy. Enough detail is included regarding each muscle's origin, insertion, action, nerve supply (including the nerve's common course or path), and blood supply to meet the requirements of the student and practitioner. This information is presented accurately and in a particularly clear and user-friendly format, especially as the specialist terminology used in anatomy can appear overwhelming at first. Technical terms are therefore explained in parentheses throughout the text.

The information about each muscle is presented in a uniform style throughout.

Attachments
A muscle is usually attached to two bones that form a joint, and when the muscle contracts, it pulls the movable bone toward the stationary bone. All muscles have at least two attachments. The origin (red) and insertion (blue) of, for example, trapezius is shown below.

Origin
The attachment that remains relatively immobile during muscular contraction. This is usually the end of the muscle which is fixed to the bone, thereby acting as an anchor for the muscle to pull its opposite end (insertion) toward this stable attachment.

Insertion
The attachment that moves, hence the opposite end of the muscle to the origin. For certain movements, when the insertion remains relatively fixed and the origin moves, the muscle is said to be performing a reversed action from the origin to insertion. Generally, the origin is more proximal (toward the center of

Origin and insertion of trapezius.

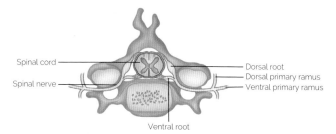

Spinal cord —
Spinal nerve —
— Dorsal root
— Dorsal primary ramus
— Ventral primary ramus
Ventral root

A spinal segment is the part of the spinal cord that gives rise to each pair of spinal nerves, one for each side of the body. Each spinal nerve contains sensory and motor fibers from the dorsal and ventral roots respectively. Soon after the spinal nerve exits through the foramen or opening between adjacent vertebrae, it divides into a dorsal primary ramus, which is directed posteriorly, and a ventral primary ramus, which is directed anteriorly and laterally. Fibers from the dorsal rami innervate the skin and extensor muscles of the neck and trunk. The ventral rami supply the limbs, as well as the sides and front of the trunk.

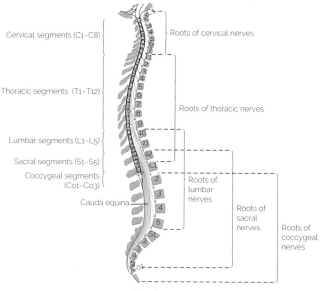

Cervical segments (C1–C8) — Roots of cervical nerves

Thoracic segments (T1–T12) — Roots of thoracic nerves

Lumbar segments (L1–L5)
Sacral segments (S1–S5)
Coccygeal segments (Co1–Co3)
Cauda equina

Roots of lumbar nerves
Roots of sacral nerves
Roots of coccygeal nerves

Spinal nerves

the body) and the insertion is more distal (toward the periphery of the body).

Nerve Supply

The *peripheral nervous system (PNS)* comprises all the neural structures outside the brain and spinal cord, which constitute the *central nervous system (CNS)*. The PNS has two main components: the *somatic nervous system* and the *autonomic nervous system*; the latter deals with involuntary control of smooth muscle and glands. As this book is concerned with skeletal muscles, it is only the somatic nervous system that is of interest.

The PNS consists of 12 pairs of cranial nerves and 31 pairs of spinal nerves, along with their subsequent branches. The spinal nerves are numbered according to the level of the spinal cord from which they arise, known as the *spinal segment*.

In this book the relevant peripheral nerve supply is listed with each muscle. However, the spinal segment from which the nerve fibers arise often varies between different sources. This is because spinal nerves are organized into networks known as *plexuses* (plexus = a network of nerves: from Latin *plectere* = "to braid"), which supply different regions of the body, and nerve fibers from different spinal segments will contribute to the individual named nerve that supplies a particular muscle.

For each muscle in this book, the spinal levels that typically contribute to its named nerve are indicated. The relevant spinal segments are represented by C for cervical, T for thoracic, L for lumbar, and S for sacral, followed by a number representing the level. Numbers in parentheses indicate a smaller contribution.

Blood Supply

When researching the arterial blood supply to each muscle, it became clear that this information is hard to come by if one is looking for clarity and consistency. Different sources sometimes disagree, especially regarding many of the smaller, deeper muscles. Other apparent contradictions merely reflect the emphasis on a different part of the arterial "chain" through which the blood travels to supply a muscle. For example, some sources will give one of the arteries supplying the rectus abdominis muscle as the superior epigastric artery, while another source will credit the blood supply to the internal thoracic artery. Because the superior epigastric artery is a branch of the internal thoracic artery, this simply indicates that one source is expressing more detail in their description than the other.

In this book I have labeled the arteries immediately supplying each muscle, but have additionally mentioned the arteries "upstream" that feed into them. Therefore, the blood vessel that actually connects with the muscle is given first and in bold, often with the blood vessel immediately upstream written in bold on the same line. The major artery that is the course of that blood vessel is then given in plain text and in parenthesis. Where applicable, I have mentioned an intermediary connecting artery if the "chain" of arteries to a muscle is extensive. For example, in the case of the iliacus muscle, the blood supply is expressed thus:

Iliolumbar branch of the internal iliac artery
via common iliac artery (from abdominal aorta).

So, if the iliolumbar branch artery is analogous to an irrigation channel that branches off from the internal iliac artery, and the internal iliac artery is itself fed from the more central abdominal aorta via the common iliac artery, we have a comprehensive overview of the route taken by the blood to reach its target.

Where a muscle is clearly supplied by a certain artery, but may also be supplied by a secondary artery (either because it applies to some people but not all, or because many but not all authorities agree that it does), I have added the potential secondary supply in plain text, as shown below:

Inferior gluteal artery
via internal iliac artery (a branch of the common iliac artery from abdominal aorta), plus can also be supplied by medial circumflex arteries (from deep femoral artery).

Where there is more than one blood supply of more or less equal importance, as is the case with the diaphragm, it is shown thus:

Musculophrenic artery
via internal thoracic artery (from subclavian artery).

Superior phrenic artery
(from thoracic aorta).

Inferior phrenic artery
(from abdominal aorta).

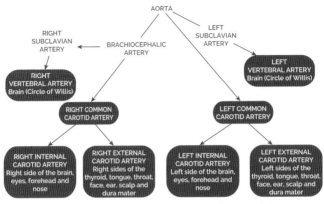

Primary blood vessels of the face, head, and neck.

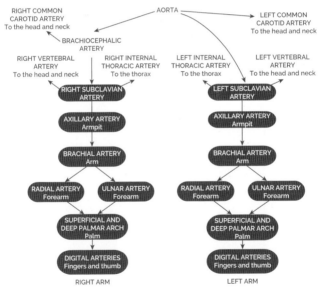

Primary blood vessels of the upper limbs.

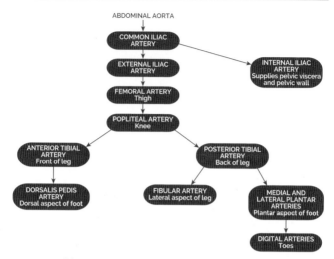

Primary blood vessels of the pelvis and lower limbs.

CHAPTER 1

Anatomical Terms

Positions

To describe the relative positions of body parts and their movements, it is essential to have a universally accepted initial reference position. This is known as the *anatomical position*, which is simply the upright standing position, with feet flat on the floor, arms hanging by the sides and the palms facing forward (see Figure 1.1). The directional

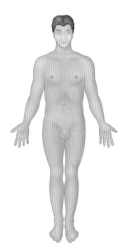

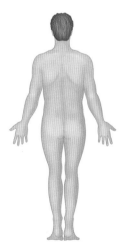

Figure 1.1: Anterior. In front of; toward or at the front of the body.

Figure 1.2: Posterior. Behind; toward or at the back of the body.

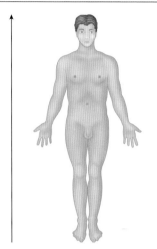

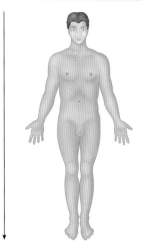

Figure 1.3: Superior. Above; toward the head or the upper part of the structure or the body.

Figure 1.4: Inferior. Below; away from the head or toward the lower part of the structure or the body.

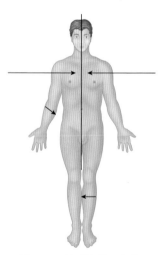

Figure 1.5: Medial. (from Latin *medius* = "middle"). Toward the midline of the body; on the inner side of a limb.

Figure 1.6: Lateral. (from Latin *latus* = "side"). Away from the midline of the body, on the outer side of the body or a limb.

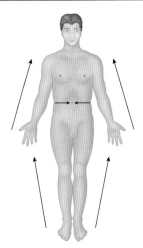

Figure 1.7: Proximal. (from Latin *proximus* = "nearest"). Closer to the center of the body (the navel), or to the point of attachment of a limb to the trunk.

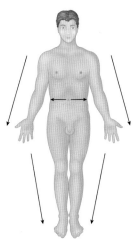

Figure 1.8: Distal. (from Latin *distans* = "distant"). Farther away from the center of the body, or from the point of attachment of a limb to the trunk.

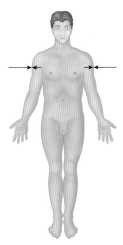

Figure 1.9: Superficial. Toward or at the body surface.

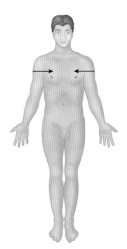

Figure 1.10: Deep. Farther away from the body surface; more internal.

Figure 1.11: Dorsal.
(from Latin *dorsum* = "back"). On the posterior surface, e.g., the back of the hand.

Figure 1.12: Palmar.
(from Latin *palma* = "palm"). On the anterior surface of the hand, i.e., the palm.

Figure 1.13: Plantar.
(from Latin *planta* = "sole"). On the sole of the foot.

terminology used always refers to the body as if it were in the anatomical position, regardless of its actual position. Note also that the terms *left* and *right* refer to the sides of the object or person being viewed, and not those of the reader.

Regions

The two primary divisions of the body are its *axial* parts, consisting of the head, neck, and trunk, and its *appendicular* parts, consisting of the limbs, which are attached to the axis of the body. Figures 1.14 and 1.15 show the terms used to indicate specific body areas. Terms in parentheses are the lay terms for the area.

Planes

The term *plane* refers to a two-dimensional section through the body; it provides a view of the body or body part, as though it has been cut through by an imaginary line.

- The sagittal planes cut vertically through the body from anterior to posterior, dividing it into right and left halves. Figure 1.16 shows the mid-sagittal plane. A *para-sagittal plane* divides the body into unequal right and left parts.
- The frontal (coronal) planes pass vertically through the body, dividing it into anterior and posterior sections, and lie at right angles to the sagittal plane.

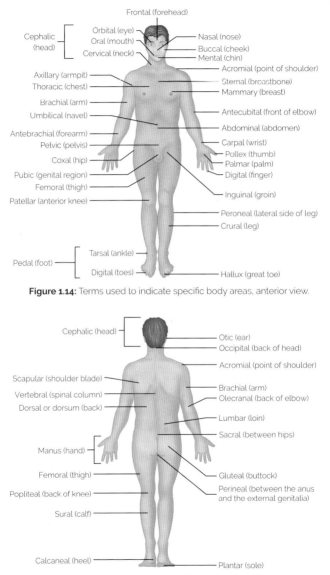

Figure 1.14: Terms used to indicate specific body areas, anterior view.

Figure 1.15: Terms used to indicate specific body areas, posterior view.

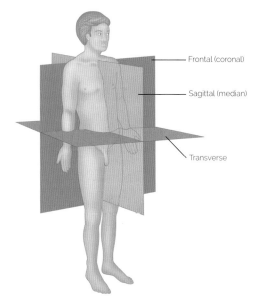

Frontal (coronal)

Sagittal (median)

Transverse

Figure 1.16: The most frequently used planes of the body.

- The transverse planes are horizontal cross sections, dividing the body into upper (superior) and lower (inferior) sections, and lie at right angles to the other two planes.

Movements

The direction in which body parts move is described in relation to the fetal position. Moving into the fetal position results from flexion of all the limbs; straightening out of the fetal position results from extension of all the limbs.

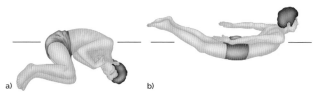

a)

b)

Figure 1.17: (a) Flexion into the fetal position. (b) Extension out of the fetal position.

Main Movements

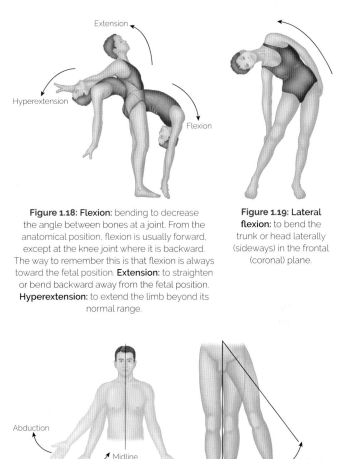

Figure 1.18: Flexion: bending to decrease the angle between bones at a joint. From the anatomical position, flexion is usually forward, except at the knee joint where it is backward. The way to remember this is that flexion is always toward the fetal position. **Extension:** to straighten or bend backward away from the fetal position. **Hyperextension:** to extend the limb beyond its normal range.

Figure 1.19: Lateral flexion: to bend the trunk or head laterally (sideways) in the frontal (coronal) plane.

Figure 1.20: Abduction: movement of a bone away from the midline of the body or a limb. **Adduction:** movement of a bone toward the midline of the body or a limb.

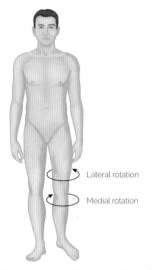

Figure 1.21: Rotation: movement of a bone or the trunk around its own longitudinal axis. **Medial or internal rotation:** to turn inward, toward the midline. **Lateral or external rotation:** to turn outward, away from the midline.

Other Movements

Movements described in this section are those that occur only at specific joints or parts of the body, usually involving more than one joint.

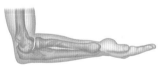

Figure 1.22: Pronation: to turn the palm of the hand down to face the floor (if standing with elbow bent 90 degrees, or if lying flat on the floor) or away from the anatomical and fetal positions.

Figure 1.23: Supination: to turn the palm of the hand up to face the ceiling (if standing with elbow bent 90 degrees, or if lying flat on the floor) or toward the anatomical and fetal positions.

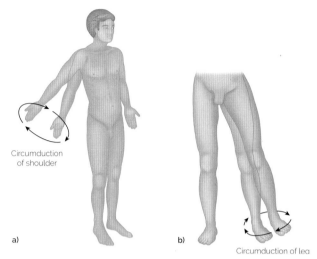

Circumduction
of shoulder

a)

b)

Circumduction of leg

Figure 1.24: Circumduction: movement in which the distal end of a bone moves in a circle, while the proximal end remains stable; the movement combines flexion, abduction, extension, and adduction.

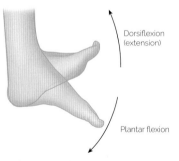

Dorsiflexion
(extension)

Plantar flexion

Figure 1.25: Plantar flexion: to point the toes down toward the ground.
Dorsiflexion: to point the toes up toward the ceiling.

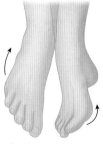

Eversion   Inversion

Figure 1.26: Inversion: to turn the sole of the foot inward, so that the soles would face toward each other. **Eversion:** to turn the sole of the foot outward, so that the soles would face away from each other.

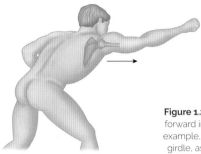

Figure 1.27: Protraction: movement forward in the transverse plane—for example, protraction of the shoulder girdle, as in rounding the shoulder.

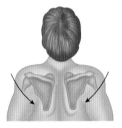

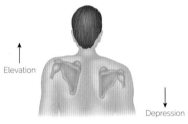

Elevation

Depression

Figure 1.28: Retraction: movement backward in the transverse plane, as in bracing the shoulder girdle back, military style.

Figure 1.29: Elevation: movement of a part of the body upward along the frontal plane—for example, elevating the scapula by shrugging the shoulders. **Depression:** movement of an elevated part of the body downward to its original position.

Opposition

Figure 1.30: Opposition: a movement specific to the saddle-shaped joint of the thumb; it enables you to touch your thumb to the tips of the fingers of the same hand.

CHAPTER 2

The Muscular System

Structure and Function of Skeletal Muscle

Skeletal (somatic or voluntary) muscles make up approximately 40% of the total human body weight. Their primary function is to produce movement through the ability to contract and relax in a coordinated manner. They are attached to bone either directly or more often via tendons. The location where a muscle attaches to a relatively stationary point on a bone, either directly or via a tendon, is called the *origin*. When the muscle contracts, it transmits tension to the bones across one or more joints, and movement occurs. The end of the muscle that attaches to the bone that moves is called the *insertion*.

Overview of Skeletal Muscle Structure

The functional unit of skeletal muscle is known as a *muscle fiber*, which is an elongated, cylindrical cell with multiple nuclei, ranging from 10 to 100 microns in width, and a few millimeters to 30+ centimeters in length. The cytoplasm of the fiber is called the *sarcoplasm*, which is encapsulated inside a cell membrane called the *sarcolemma*. A delicate membrane known as the *endomysium* surrounds each individual fiber.

The muscle fibers are grouped together in bundles covered by a collagenic sheath known as the *perimysium*. These bundles are themselves grouped together, and the whole muscle is encased in a sheath called the *epimysium*. These muscle membranes lie throughout the entire length of the muscle, from the tendon of origin to the tendon of the insertion. This entire structure is sometimes referred to as the *musculotendinous unit*.

In defining the structure of muscle tissue in more detail, from microscopic to gross anatomy, we therefore have the following components: myofibrils, endomysium, fasciculi, perimysium, epimysium, deep fascia, and superficial fascia.

Myofibrils

Through an electron microscope, one can distinguish the contractile elements of a muscle fiber, known as *myofibrils*, running the entire length of the fiber. Each myofibril reveals alternate light and dark banding, producing the characteristic cross-striation of the muscle fiber; these bands are called *myofilaments*. The light bands are referred to as *isotropic (I) bands*, and consist of thin myofilaments made of the protein actin. The dark ones are called *anisotropic (A) bands*, consisting of thicker myofilaments made of the protein myosin. A third connecting filament is made of the sticky protein titin, also known as connectin, which is the third most abundant protein in human tissue.

The myosin filaments have paddle-like extensions that emanate from them, rather like the oars of a boat. These extensions latch onto the actin filaments, forming what are described as *cross-bridges* between the two types of filaments. These cross-bridges, using the muscle energy source known as *adenosine triphosphate (ATP)*, pull the actin strands closer together.* Thus, the light and dark sets of filaments increasingly overlap, like interlocking fingers, resulting in muscle contraction. A set of actin-myosin filaments is called a *sarcomere*.

- The lighter zone is known as the *I band*, and the darker zone the *A band*.
- The *Z line* is a thin dark line at the midpoint of the I band.
- A *sarcomere* is defined as the section of myofibril between consecutive Z lines.
- The center of the A band contains the *H zone*.
- The *M line* bisects the H zone, and delineates the center of the sarcomere.

---

* **Huxley's Sliding Filament Theory**. The generally accepted hypothesis to explain muscle function is partly described by the sliding filament theory proposed by Huxley and Hanson in 1954. Muscle fibers receive a nerve impulse that causes the release of calcium ions stored in the muscle. In the presence of the ATP, the calcium ions bind with the actin and myosin filaments to form an electrostatic (magnetic) bond. This bond causes the fibers to shorten, resulting in their contraction or an increase in tonus (muscle tone). When the nerve impulse ceases, the muscle fibers relax. Because of their elastic nature, the filaments recoil to their non-contracted lengths, i.e., their resting level of tonus.

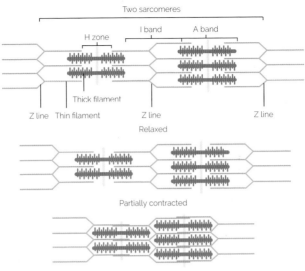

Figure 2.1: Filaments within a muscle fibre move to overlap one another and thereby cause the fibre to shorten or contract.

If an outside force causes a muscle to stretch beyond its resting level of tonus (the slight, continuous contraction of a muscle, aiding the maintenance of posture), the interlinking effect of the actin and myosin filaments that occurs during contraction is reversed. Initially, the actin and myosin filaments accommodate the stretch, but as the stretch continues, the titin filaments increasingly "pay out" to absorb the displacement. Thus, it is the titin filament that determines the muscle fiber's extensibility and resistance to stretch.

Research indicates that a muscle fiber (sarcomere) can be elongated to 150% of its normal length at rest.

Endomysium
A delicate connective tissue called *endomysium* lies outside the sarcolemma of each muscle fiber, separating each fiber from its neighbors, but also connecting them together.

Fasciculi
Muscle fibers are arranged in parallel bundles called *fasciculi*.

Perimysium

Each fasciculus is bound by a denser collagenic sheath called the *perimysium*.

Epimysium

The entire muscle, which is therefore an assembly of fasciculi, is wrapped in a fibrous sheath called the *epimysium*; this arrangement facilitates force transmission.

Deep Fascia

A coarser sheet of fibrous connective tissue lies outside the epimysium, binding individual muscles into functional groups. This deep fascia extends to wrap around other adjacent structures.

Superficial Fascia

While its anatomy and topography differs from region to region, providing specialization, the superficial fascia is primarily a fatty layer that contains oblique septa and connects the skin to deep fascia. Contractile fibers have been reported in the superficial fascia, particularly in the neck.

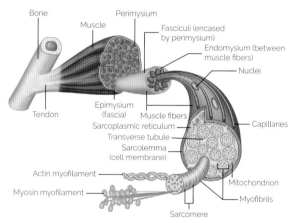

Figure 2.2: The structure of muscle tissue from microscopic to gross anatomy.

Muscle Attachment

The way a muscle attaches to bone or other tissues is through either a direct or an indirect attachment. A *direct* or *fleshy attachment* is where the perimysium and epimysium of the muscle unite and fuse with the periosteum of bone, perichondrium of cartilage, a joint capsule, or the connective tissue underlying the skin, as in some muscles of facial expression. An *indirect attachment* is where the connective tissue components of a muscle fuse together into bundles of collagen fibers to form an intervening tendon. Indirect attachments are much more common. The different types of indirect attachment are: tendons and aponeuroses, intermuscular septa, and sesamoid bones.

Tendons and Aponeuroses

When the connective tissue components of a muscle combine and extend beyond the end of the muscle as round cords or flat bands, the tendinous attachment is called a *tendon*; if they extend as a thin, flat, and broad sheet-like material, the attachment is called an *aponeurosis*. The tendon or aponeurosis secures the muscle to bone or cartilage, to the fascia of other muscles, or to a seam of fibrous tissue called a *raphé*. Flat patches of tendon may form on the body of a muscle where it is exposed to friction. This may occur, for example, on the deep surface of the trapezius where it rubs against the spine of the scapula.

Intermuscular Septa

In some cases, flat sheets of dense connective tissue known as *intermuscular septa* penetrate between muscles, providing another structure to which muscle fibers may attach.

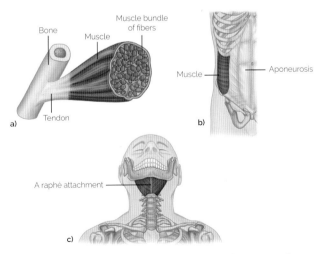

Figure 2.3: (a) Tendon attachment; (b) attachment by aponeurosis; (c) mylohyoid raphé.

Sesamoid Bones

If a tendon is subject to friction, it may, though not in all cases, develop a sesamoid bone within its substance. The largest sesamoid bone in the body is the patella or kneecap. However, sesamoid bones may also appear in tendons not subject to friction.

Multiple Attachments

Many muscles have only two attachments, one at each end. More complex muscles, on the other hand, are often attached to several different structures at their origins and/or their insertions. If these attachments are separated, so that there are two or more tendons and/or aponeuroses inserting into different places, the muscle is said to have two or more heads. For example, the biceps brachii has two heads at its origin: one from the coracoid process of the scapula, and the other from the supraglenoid tubercle. The triceps brachii has three heads and the quadriceps femoris has four.

Red and White Muscle Fibers

Three types of skeletal muscle fibers have been distinguished: (1) red slow-twitch fibers; (2) white fast-twitch fibers; and (3) intermediate fast-twitch fibers. There is always a mixture of these types of muscle fibers in any given muscle, giving them a range of fatigue resistance and contractile speeds.

1. *Red slow-twitch fibers*: These fibers are thin cells that contract slowly. The red color is due to their content of myoglobin, a substance similar to hemoglobin, which stores oxygen and increases the rate of oxygen diffusion within the muscle fibers. As long as the oxygen supply is plentiful, red fibers can contract for sustained periods, and are thus very resistant to fatigue. Successful marathon runners tend to have a high percentage of these red fibers.
2. *White fast-twitch fibers*: These fibers are large cells that contract rapidly. They are pale because of their lower content of myoglobin. White fibers fatigue quickly, because they rely on short-lived glycogen reserves in the fiber to contract. However, they are capable of generating much more powerful contractions than red fibers, enabling them to perform rapid, powerful movements for short periods. Successful sprinters have a higher proportion of these white fibers.
3. *Intermediate Fast-twitch fibers*: These red or pink fibers are a compromise in size and activity between red and white fibers.

Blood Supply

In general, each muscle receives an arterial supply to bring nutrients via the blood into the muscle, and contains several veins to take away metabolic by-products released by the muscle into the blood. These blood vessels usually enter through the central part of the muscle, but may also enter toward one end. Thereafter, they branch into a capillary plexus, which spreads throughout the intermuscular septa, to eventually penetrate the endomysium around each muscle fiber. During exercise, the capillaries dilate, increasing the amount of blood flow in the muscle by up to 800 times. However, a muscle tendon, being composed of a relatively inactive tissue, has a much less extensive blood supply.

Nerve Supply

The nerve supply to a muscle usually enters at the same place as the blood supply in a neurovascular bundle, and branches through the connective tissue septa into the endomysium in a similar way. Each skeletal muscle fiber is supplied by a single nerve ending. This is in contrast to some other muscle tissues, which are able to contract without any nerve stimulation.

The nerve entering the muscle usually contains roughly equal proportions of sensory and motor nerve fibers, although some muscles may receive separate sensory branches. As the nerve fiber approaches the muscle fiber, it divides into a number of terminal branches, collectively known as a *motor end plate*.

Motor Unit of a Skeletal Muscle

A motor unit consists of a single motor nerve cell and the muscle fibers that it stimulates. Motor units vary in size, ranging from cylinders of muscle 5–7 mm in diameter in the upper limb, to 7–10 mm in diameter in the lower limb. The average number of muscle fibers within a unit is 150, but this number can range from less than ten to several hundred. Where fine gradations of movement are required, as in the muscles of the eyeball or fingers, the number of muscle fibers supplied by a single nerve cell is small. On the other hand, where mass movements are required, as in the muscles of the lower limb, each nerve cell may supply a motor unit of several hundred fibers.

The muscle fibers in a single motor unit are spread throughout the muscle, rather than being clustered together. This means that stimulation of a single motor unit will cause the entire muscle to exhibit a weak contraction.

Skeletal muscles work on an "all or nothing principle": in other words, groups of muscle cells, or fasciculi, can either contract or not at all. Depending on the strength of contraction required, a certain number of muscle cells will fully contract, while others will not. When a greater muscular effort is needed, most of the motor units may be stimulated at the same time. However, under normal conditions, the motor units tend to work in relays, so that during prolonged contractions some are inhibited while others are contracting—this is known as *gradual increments of contraction (GIC)*.

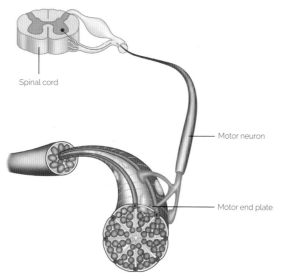

Figure 2.4: A motor unit of a skeletal muscle.

Muscle Reflexes
Within skeletal muscles there are two specialized types of nerve receptor that can sense tension (length or stretch): muscle spindles and Golgi tendon organs (GTOs). *Muscle spindles* are cigar-like in shape and consist of tiny modified muscle fibers called *intrafusal fibers*, and nerve endings, encased together within a connective tissue sheath; they lie between and parallel to the main muscle fibers.

GTOs are located mostly at the junctions of muscles and their tendons or aponeuroses.

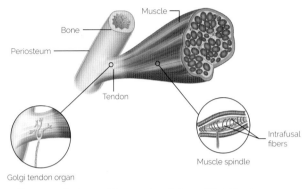

Figure 2.5: Anatomy of the muscle spindle and Golgi tendon organ.

Stretch Reflex

The *stretch reflex* helps control posture by maintaining muscle tone. It also helps prevent injury, by enabling a muscle to respond to a sudden or unexpected increase in length. This is how it works:

1. When a muscle is lengthened, the muscle spindles are excited, causing each spindle to send a nerve impulse communicating the speed of lengthening to the spinal cord.
2. On receiving this impulse, the spinal cord immediately sends a proportionate impulse back to the stretched muscle fibers, causing them to contract, in order to decelerate the movement. This circular process is known as the *reflex arc*.
3. An impulse is simultaneously sent from the spinal cord to the antagonist of the contracting muscle (i.e., the muscle opposing the contraction), inhibiting the action of the antagonist so that it cannot resist the contraction of the stretched muscle. This process is known as *reciprocal inhibition*.
4. Concurrent with this spinal reflex, nerve impulses are also sent up the spinal cord to the brain to relay information about muscle length and the speed of muscle contraction. A reflex in the brain feeds nerve impulses back to the muscles in order to ensure that appropriate muscle tone is maintained to meet the requirements of posture and movement.

5.   Meanwhile, the stretch sensitivity of the minute intrafusal muscle fibers within the muscle spindle are evened out and regulated by gamma efferent nerve fibers*, arising from motor neurons within the spinal cord. Thus, a gamma motor neuron reflex arc ensures the evenness of muscle contraction, which would otherwise be jerky if muscle tone relied solely on the stretch reflex.

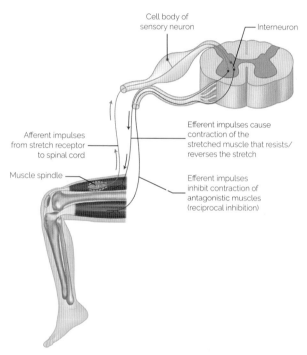

Cell body of
sensory neuron

Interneuron

Afferent impulses
from stretch receptor
to spinal cord

Muscle spindle

Efferent impulses cause
contraction of the
stretched muscle that resists/
reverses the stretch

Efferent impulses
inhibit contraction of
antagonistic muscles
(reciprocal inhibition)

Figure 2.6: The stretch reflex arc.

The classic clinical example of the stretch reflex used in clinical practice is the *knee jerk*, or *patellar reflex*, whereby the patellar ligament is lightly struck with a small rubber hammer. This results in the following sequence of events:

---

*The function of these nerve fibers is to regulate the sensitivity of the spindle and the total tension in the muscle

1. The sudden stretch of the patellar ligament causes the quadriceps to be stretched, i.e., the sharp tap on the patellar ligament causes a sudden stretch of the tendon.
2. This rapid stretch is registered by the muscle spindles within the quadriceps, causing the quadriceps to contract. This causes a small kick as the knee straightens suddenly, and takes the tension off the muscle spindles.
3. Simultaneously, nerve impulses to the hamstrings, which are the antagonists of the quadriceps, result in functional inhibition of their action.

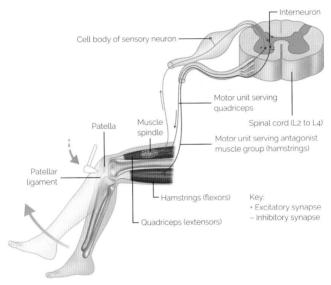

Figure 2.7: The patellar reflex.

Another well-known example of the stretch reflex in action occurs when a person falls asleep in the seated position: their head will relax forward, then jerk back up, because the stretched muscle spindles in the back of the neck have activated a reflex arc.

The stretch reflex also works constantly to maintain the tonus of our postural muscles; in other words, it enables us to remain standing without conscious effort and without collapsing forward.

The sequence of events preventing this forward collapse occurs in a fraction of a second, as follows:

1. In standing, we naturally begin to sway forward.
2. This pulls our calf muscles into a lengthened position, activating the stretch reflex.
3. The calf muscles consequently contract to pull us back to the upright position.

Deep Tendon Reflex (Autogenic Inhibition)

In contrast to the stretch reflex, which involves the response of the muscle spindles to elongation of muscle fibers, the *deep tendon reflex* involves the reaction of GTOs to muscle contraction or an undue rise in tension. Accordingly, the deep tendon reflex creates the opposite effect to that of the stretch reflex. This is how it works:

1. When a muscle contracts, it pulls on the tendons which are situated at both ends.
2. The tension in the tendon causes the GTOs to transmit impulses to the spinal cord. Some impulses continue to the cerebellum.
3. As these impulses reach the spinal cord, they inhibit the motor nerves supplying the contracting muscle, thus reducing tonus.
4. Simultaneously, the motor nerves supplying the antagonist muscle are activated, causing it to contract. This process is called *reciprocal activation*.
5. Meanwhile, the information reaching the cerebellum is processed and fed back to help readjust muscle tension.

The deep tendon reflex has a protective function: it prevents the muscle from contracting so hard that it would pull its attachment off the bone. It is therefore especially important during activities which involve rapid switching between flexion and extension, such as running.

Note, however, that in normal day-to-day movement, tension in the muscles is not sufficient to activate the GTOs and cause a deep tendon reflex. By contrast, the threshold of the muscle spindle stretch reflex is set much lower, because it must constantly maintain sufficient tonus in the postural muscles to keep the body upright.

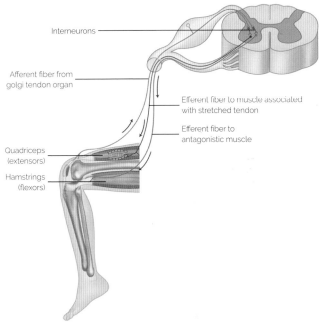

Figure 2.8: The deep tendon reflex.

Isometric and Isotonic Contractions

A muscle will contract upon stimulation in an attempt to bring its attachments closer together, but this does not necessarily result in a shortening of the muscle. If the contraction of a muscle results in no movement, such a contraction is called *isometric*; if movement of some sort results, the contraction is called *isotonic*.

Isometric Contraction

An *isometric* contraction occurs when there is increased tension in a muscle, but its length remains unchanged. In other words, although the muscle tenses, the joint over which the muscle passes does not move. One example of this is holding a heavy object in the hand with the elbow held stationary and bent at 90 degrees. Trying to lift something that proves to be too heavy to move is another example.

Note also that some of the postural muscles are largely working isometrically by automatic reflex. For example, in the upright position, the body has a natural tendency to fall forward at the ankle; this is prevented by isometric contraction of the calf muscles. Likewise, the center of gravity of the skull would make the head tilt forward if the muscles at the back of the neck did not contract isometrically to keep the head centralized.

Isotonic Contraction

Isotonic contractions of muscle enable us to move about.
Such contractions are of two types: concentric and eccentric.

In *concentric* contractions, the muscle attachments move closer together, causing movement at the joint. In the example of holding an object in the hand, if the biceps muscle contracts concentrically, the elbow joint will flex and the hand will move toward the shoulder. Similarly, if we look up at the ceiling, the muscles at the back of the neck must contract concentrically to tilt the head back and extend the neck.

Eccentric contraction means that the muscle fibers "pay out" in a controlled manner to slow down movements in a case where gravity, if unchecked, would otherwise cause them to occur too rapidly, as, for example, when lowering an object held in the hand down to your side. Another everyday example is simply sitting down onto a chair. Therefore, the difference between concentric and eccentric contractions is that in the former, the muscle shortens, while in the latter, it actually lengthens.

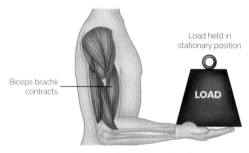

Load held in
stationary position

Biceps brachii
contracts

LOAD

Figure 2.9: Isometric contraction.

Figure 2.10: Abdominal muscles contract concentrically to raise the body.

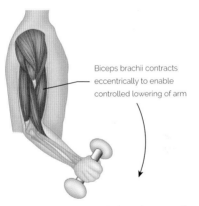

Biceps brachii contracts eccentrically to enable controlled lowering of arm

Figure 2.11: Eccentric isotonic contraction.

Muscle Shape (Arrangement of Fascicles)

Muscles come in a variety of shapes according to the arrangement of their fascicles. The reason for this variation is to provide optimum mechanical efficiency for a muscle in relation to its position and action. The most common arrangement of fascicles yields muscle shapes which can be described as parallel, pennate, convergent, and circular, with each of these shapes having further sub-categories. The different shapes are illustrated in Figure 2.12.

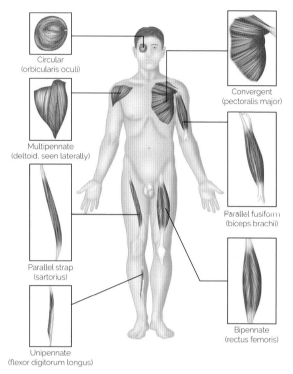

Circular
(orbicularis oculi)

Convergent
(pectoralis major)

Multipennate
(deltoid, seen laterally)

Parallel fusiform
(biceps brachii)

Parallel strap
(sartorius)

Bipennate
(rectus femoris)

Unipennate
(flexor digitorum longus)

Figure 2.12: Muscle shapes.

Parallel

In this arrangement the fascicles are arranged parallel to the long
axis of the muscle. If the fascicles extend throughout the length of the
muscle, it is known as a *strap muscle*, as, for example, the sartorius.
If the muscle also has an expanded belly and tendons at both ends,
it is called a *fusiform muscle*, as, for example, the biceps brachii.
A variation of this type of muscle has a fleshy belly at either end,
with a tendon in the middle; as in the digastric muscle.

Pennate

Pennate muscles are so named because their short fasciculi are attached
obliquely to the tendon, like the structure of a feather (from Latin
penna = "feather"). If the tendon develops on one side of the muscle,
it is referred to as *unipennate*, as in, for example, the flexor digitorum

longus in the leg. If the tendon is in the middle and the fibers are attached obliquely from both sides, it is known as *bipennate*, a good example being the rectus femoris. If there are numerous tendinous intrusions into the muscle, with fibers attaching obliquely from several directions (thus resembling many feathers side by side), the muscle is referred to as *multipennate*; the best example is the deltoid muscle.

Convergent
Muscles that have a broad origin with fascicles converging toward a single tendon, giving the muscle a triangular shape, are called *convergent muscles*. The best example is the pectoralis major.

Circular
When the fascicles of a muscle are arranged in concentric rings, the muscle is referred to as *circular*. All the sphincter skeletal muscles in the body are of this type; they surround openings, which they close by contracting. An example is the orbicularis oculi.

Range of Motion Versus Power
When a muscle contracts, it can shorten by up to 70% of its original length; hence, the longer the fibers, the greater the range of motion. On the other hand, the strength of a muscle depends on the total number of muscle fibers it contains, rather than the length of the fibers. Therefore:

1. Muscles with long parallel fibers produce the greatest range of motion, but are not usually very powerful.
2. Muscles with a pennate pattern (especially multipennate) pack in the most fibers. Such muscles shorten less than long parallel muscles, but tend to be much more powerful.

Functional Characteristics of a Skeletal Muscle
All that has been said about muscles so far in this book enables us to formulate a list of functional characteristics pertaining to skeletal muscle.

Excitability
Excitability is the ability to receive and to respond to a stimulus. In the case of a muscle, when a nerve impulse from the brain reaches the muscle, a chemical known as *acetylcholine* is released. This chemical produces a change in the electrical balance in the muscle fiber and, as a result, generates an electrical current known as an *action potential*. The action potential conducts the electrical current from one end of the muscle cell to the other and results in a contraction of the muscle cell, or muscle fiber (remember that one muscle cell = one muscle fiber).

Contractility

Contractility is the ability of a muscle to shorten forcibly when stimulated. The muscles themselves can only contract; they do not usually lengthen, except via some external means (i.e., manually), beyond their normal resting length (see "Tonus" below). In other words, muscles can only pull their ends together (contract); they cannot push them apart.

Extensibility

Extensibility is the ability of a muscle to be extended, or returned to its resting length (which is a semi-contracted state) or slightly beyond. For example, if we bend forward at the hips from standing, the muscles of the back, such as the erector spinae, lengthen eccentrically (see p. 137) to lower the trunk, paying out slightly beyond their normal resting length, and are thus effectively "elongated."

Elasticity

Elasticity describes the ability of a muscle fiber to recoil after being lengthened, and therefore resume its resting length when relaxed. In a whole muscle, the elastic effect is supplemented by the important elastic properties of the connective tissue sheaths (endomysium and epimysium). Tendons also contribute some elastic properties. An example of this elastic recoil effect can be experienced when coming back up from a forward bend at the hips as described above. Initially there is no muscle contraction; instead, the upward movement is initiated purely by elastic recoil of the back muscles, after which the contraction of the back muscles completes the movement.

Tonus

Tonus, or *muscle tone*, is the term used to describe the slightly contracted state to which muscles return during the resting state. Muscle tonus does not produce active movements, but it keeps the muscles firm, healthy, and ready to respond to stimulation. It is the tonus of skeletal muscles that also helps stabilize and maintain posture. *Hypertonic muscles* are those muscles whose "normal" resting state is over-contracted.

General Functions of Skeletal Muscles

- **Enable movement:** skeletal muscles are responsible for all locomotion and manipulation, and they enable you to respond quickly.
- **Maintain posture:** skeletal muscles support an upright posture against the pull of gravity.
- **Stabilize joints:** skeletal muscles and their tendons stabilize joints.

- **Generate heat:** skeletal muscles (in common with smooth and cardiac muscles) generate heat, which is important in maintaining a normal body temperature.

Musculoskeletal Mechanics

Origins and Insertions

In the majority of movements, one attachment of a muscle remains relatively stationary while the attachment at the other end moves. The more stationary attachment is called the *origin* of the muscle, and the other attachment, the *insertion*. A spring that closes a gate could be said to have its origin on the gatepost and its insertion on the gate itself.

In the body, however, the attachment arrangement is rarely so clear-cut, because, depending on the activity one is engaged in, the fixed and movable ends of the muscle may be reversed. For example, muscles that attach the upper limb to the chest normally move the arm relative to the trunk, which means their origins are on the trunk and their insertions are on the upper limb. However, in climbing, the arms are fixed, while the trunk moves as it is pulled up to the fixed limbs. In this type of situation, where the insertion is fixed and the origin moves, the muscle is said to perform a *reversed action*. Because there are so many situations where muscles are working with a reversed action, it is sometimes less confusing to simply speak of attachments, without reference to origin and insertion.

In practice, a muscle attachment that lies more proximally (more toward the trunk or on the trunk) is usually referred to as the *origin*. An attachment that lies more distally (away from the attached end of a limb or away from the trunk) is referred to as the *insertion*.

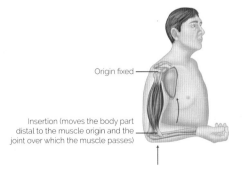

Origin fixed

Insertion (moves the body part distal to the muscle origin and the joint over which the muscle passes)

Figure 2.13: A muscle working with its origin fixed and its insertion moving.

Group Action of Muscles

Muscles work together or in opposition in order to achieve a wide variety of movements; therefore, whatever one muscle can do, there is another muscle that can undo it. Muscles may also be required to provide additional support or stability to enable certain movements to occur elsewhere.

Muscles are classified into four functional groups:

- Prime mover, or agonist
- Antagonist
- Synergist
- Fixator

Prime Mover, or Agonist

A *prime mover* (also called an *agonist*) is a muscle that contracts to produce a specific movement. An example is the biceps brachii, which is the prime mover in elbow flexion. Other muscles may assist the prime mover in providing the same movement, albeit with less effect: such muscles are called *assistant* or *secondary movers*. For example, the brachialis assists the biceps brachii in flexing the elbow, and is therefore a secondary mover.

Antagonist

The muscle on the opposite side of a joint to the prime mover, and which must relax to allow the prime mover to contract, is called an *antagonist*. For example, when the biceps brachii on the front of the arm contracts to flex the elbow, the triceps brachii on the back of the arm must relax to allow this movement to occur. When the movement is reversed (i.e., the elbow is extended), the triceps brachii becomes the prime mover and the biceps brachii assumes the role of antagonist.

Synergist

Synergists prevent any unwanted movements that might occur as the prime mover contracts. This is especially important where a prime mover crosses two joints, because when it contracts it will cause movement at both joints, unless other muscles act to stabilize one of the joints. For example, the muscles that flex the fingers not only cross the finger joints, but also cross the wrist joint, potentially causing movement at both joints. However, because you have other muscles acting synergistically to stabilize the wrist joint, you are able to flex the fingers into a fist without also flexing the wrist at the same time.

A prime mover may have more than one action, and so synergists also act to eliminate the unwanted movements. For example, the biceps brachii will flex the elbow, but its line of pull will also supinate the forearm (twist the forearm, as in tightening a screw). If you want flexion to occur without supination, other muscles must contract to prevent this supination. In this context, such synergists are sometimes called *neutralizers*.

Fixator

A synergist is more specifically referred to as a *fixator* or *stabilizer* when it immobilizes the bone from which the prime mover takes origin, thus providing a stable base for the action of the prime mover. The muscles that stabilize (fix) the scapula during movements of the upper limb are good examples. The sit-up exercise is another good example. The abdominal muscles attach to both the ribcage and the pelvis; when they contract to enable you to perform a sit-up, the hip flexors will contract synergistically as fixators to prevent the abdominals tilting the pelvis, thereby enabling the upper body to curl forward as the pelvis remains stationary.

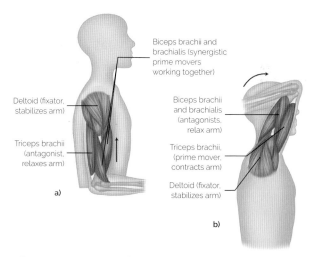

Biceps brachii and brachialis (synergistic prime movers working together)

Deltoid (fixator, stabilizes arm)

Triceps brachii (antagonist, relaxes arm)

a)

Biceps brachii and brachialis (antagonists, relax arm)

Triceps brachii, (prime mover, contracts arm)

Deltoid (fixator, stabilizes arm)

b)

Figure 2.14: Group action of muscles: (a) flexing the arm at the elbow; (b) extending the arm at the elbow (showing reversed roles of prime mover and antagonist).

Leverage

In classical biomechanics, the bones, joints, and muscles together form a system of levers in the body to optimize the relative strength, range, and speed required of any given movement. The joints act as fulcrums, the muscles apply the effort and the bones bear the weight of the body part to be moved.

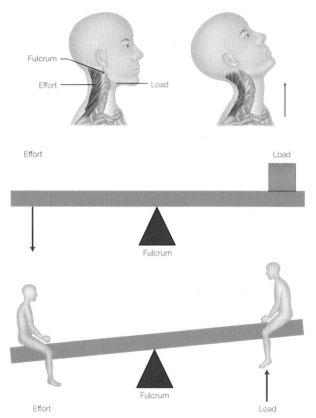

Figure 2.15: First-class lever: the relative position of the components is load–fulcrum–effort. Examples are a seesaw and a pair of scissors. In the body, an example is the ability to extend the head and neck: here the facial structures are the load, the atlanto-occipital joint is the fulcrum, and the posterior neck muscles provide the effort.

Figure 2.16: Second-class lever: the relative position of components is fulcrum–load–effort. The best example is a wheelbarrow. In the body, an example is the ability to raise the heels off the ground in standing: here the ball of the foot is the fulcrum, the body weight is the load, and the calf muscles provide the effort. With second-class levers, speed and range of movement are sacrificed for strength.

Figure 2.17: Third-class lever: the relative position of components is load–effort–fulcrum. A pair of tweezers is an example of this. In the body, most skeletal muscles act in this way. An example is flexing the forearm: here an object held in the hand is the load, the biceps brachii provides the effort, and the elbow joint is the fulcrum. With third-class levers, strength is sacrificed for speed and range of movement.

A muscle attached close to the fulcrum will be weaker than it would be if it were attached further away. However, it is able to produce a greater range and speed of movement, because the length of the lever amplifies the distance travelled by its movable attachment. Figure 2.18 illustrates this principle in relation to the adductors of the hip joint. The muscle so positioned to move the greater load (in this case the adductor longus) is said to have a *mechanical advantage*. The muscle attached close to the fulcrum is said to operate at a *mechanical disadvantage*, and has a weaker action.

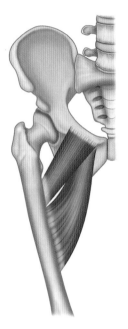

Figure 2.18: The pectineus is attached closer to the axis of movement than the adductor longus. Therefore, the pectineus is the weaker adductor of the hip, but is able to produce a greater movement of the lower limb per centimeter of contraction.

CHAPTER 3

Muscles of the Scalp and Face

The muscles of the face are unique among the muscles of the body, because most muscles connect bone to bone, whereas the facial muscles generally connect bone to skin. These muscles perform a multitude of tasks, including movement of the head and neck, chewing and swallowing, speech, facial expressions, and movement of the eyes. Such numerous and varied movements require the fastest, finest, and most delicate adjustments in the entire human body.

There are two main muscle groups involved, namely the muscles of mastication and the muscles of facial expression.

Muscles of Mastication

The muscles of mastication include masseter, temporalis, and pterygoids; these muscles work to elevate the mandible relative to the rest of the skull, closing the mouth to bite, chew, and speak.

Muscles of Facial Expression

The muscles of facial expression are very superficial muscles to allow the movement of the skin and superficial fascia in various directions. There are two major muscle groups of facial expression, namely the muscles around the eye and the muscles around the mouth.

There are many other muscles of the scalp and face that assist in facial expression, such as occipitofrontalis (scalp), the muscles of the nose, and the muscles of the external part of the ear.

Finally, the **muscles of the scalp** comprise **occipitofrontalis** and **temporoparietalis**. The former plays an important role in facial expression, such as raising the eyebrows, and the latter in tightening the scalp and raising the ears.

Muscles of the Scalp

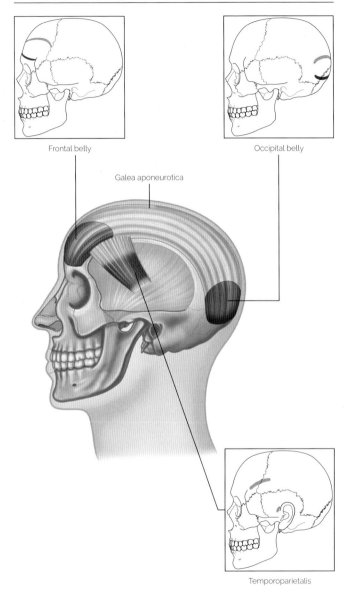

Frontal belly

Occipital belly

Galea aponeurotica

Temporoparietalis

Occipitofrontalis

Latin, *frons*, forehead, front of the head; *occiput*, back of the head.

Origin
Frontal belly
Skin of eyebrows.
Occipital belly
Lateral two-thirds of superior nuchal line of occipital bone. Mastoid process of temporal bone.

Insertion
Galea aponeurotica.

Nerve supply
Facial nerve (VII) (posterior auricular and temporal branches).

Blood supply
Frontal belly
Supraorbital and supratrochlear branches of ophthalmic artery (from internal carotid artery)
Occipital belly
Occipital artery (from external carotid artery).

Action
Frontal belly
Raises eyebrows and wrinkles skin of forehead horizontally.
Occipital belly
Pulls scalp backward.

Temporoparietalis

Latin, *tempus*, temple; *parietalis*, relating to the walls of a cavity.

Origin
Fascia above ear.

Insertion
Lateral border of galea aponeurotica.

Nerve supply
Facial nerve (VII) (temporal branch).

Blood supply
Superficial temporal and posterior auricular arteries via external carotid artery (from common carotid artery).

Action
Tightens scalp. Raises ears.

Muscles of the Ear

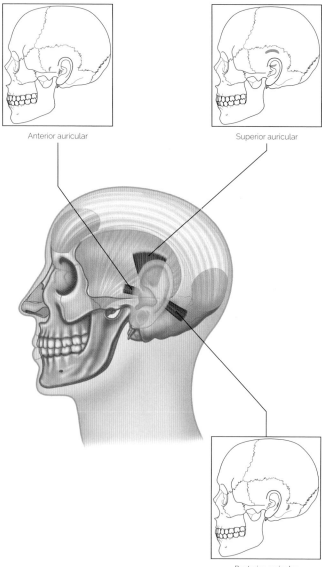

Anterior auricular

Superior auricular

Posterior auricular

Superior Auricular

Latin, *auricularis*, relating to the ear; *superior*, upper.

Origin
Fascia in temporal region above ear.

Insertion
Superior part of ear.

Nerve supply
Facial nerve (VII) (temporal branch).

Blood supply
Superficial temporal and posterior auricular arteries
via external carotid artery (from common carotid artery).

Action
Elevates ear.

Anterior Auricular

Latin, *auricularis*, relating to the ear; *anterior*, at the front.

Origin
Anterior part of temporal fascia.

Insertion
Into helix of ear.

Nerve supply
Facial nerve (VII) (temporal branch).

Blood supply
Superficial temporal and posterior auricular arteries
via external carotid artery (from common carotid artery).

Action
Draws ear forward and upward.

Posterior Auricular

Latin, *auricularis*, relating to the ear; *posterior*, at the back.

Origin
Mastoid process of temporal bone.

Insertion
Posterior part of ear.

Nerve supply
Facial nerve (VII) (posterior auricular branch).

Blood supply
Superficial temporal and posterior auricular arteries
via external carotid artery (from common carotid artery).

Action
Pulls ear backward and upward.

Muscles of the Eyelids

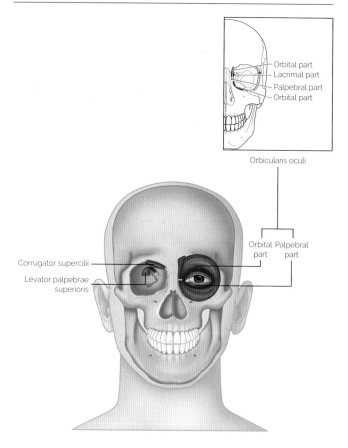

Orbital part
Lacrimal part
Palpebral part
Orbital part

Orbicularis oculi

Orbital part Palpebral part

Corrugator supercilii

Levator palpebrae superioris

Orbicularis Oculi

Latin, *orbiculus*, small circular disc; *oculus*, eye.

Origin
Orbital part
Frontal bone. Frontal process of maxilla. Medial palpebral ligament.
Palpebral part
Medial palpebral ligament.
Lacrimal part
Lacrimal bone.

Insertion
Orbital part
Circular path around orbit, returning to origin.
Palpebral part
Lateral palpebral raphe.
Lacrimal part
Lateral palpebral raphe.

Nerve supply
Facial nerve (VII) (temporal and zygomatic branches).

Blood supply
Orbital and palpebral parts
Upper fibers
Supraorbital and supratrochlear branches of ophthalmic artery
(from internal carotid artery).
Lower fibers
Infraorbital branch of maxillary artery and angular branch of facial artery
(from external carotid artery).

Lacrimal part
Infraorbital branch of maxillary artery and angular branch of facial artery
(from external carotid artery).

Action
Orbital part
Strongly closes eyelids (firmly "screws up" eye).
Palpebral part
Gently closes eyelids (and comes into action involuntarily, as in blinking).
Lacrimal part
Dilates lacrimal sac and brings lacrimal canals onto surface of eye.

Muscles of the Eyelids

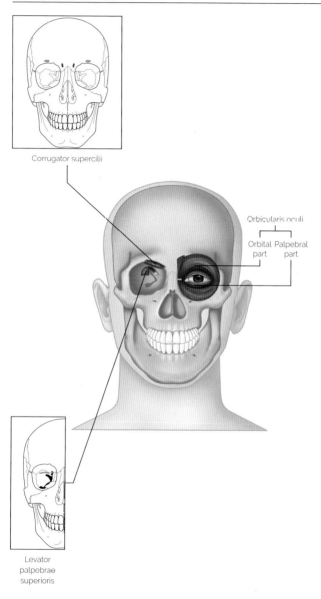

Corrugator supercilii

Orbicularis oculi

Orbital part   Palpebral part

Levator palpebrae superioris

Levator Palpebrae Superioris

Latin, *levare*, to lift; *palpebrae*, of the eyelid; *superioris*, of the upper.

Origin
Root of orbit (lesser wing of sphenoid bone).

Insertion
Skin of upper eyelid.

Nerve supply
Oculomotor nerve (III) (superior branch).

Blood supply
Ophthalmic artery
(from internal carotid artery).

Action
Raises upper eyelid.

Corrugator Supercilii

Latin, *corrugare*, to wrinkle up; *supercilii*, of the eyebrow.

Origin
Medial end of superciliary arch of frontal bone.

Insertion
Deep surface of skin under medial half of eyebrows.

Nerve supply
Facial nerve (VII) (temporal branch).

Blood supply
Supraorbital branch of ophthalmic artery
(from internal carotid artery).

Action
Draws eyebrows medially and downward, thus producing vertical wrinkles.

Muscles of the Nose

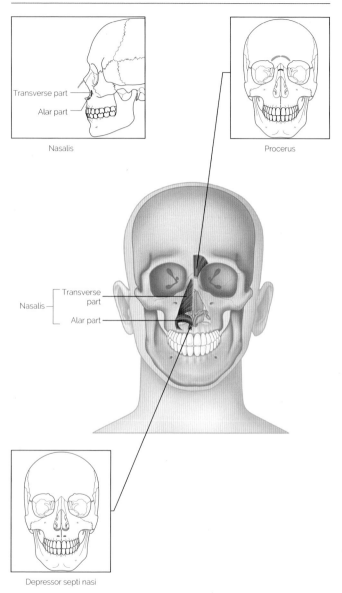

Transverse part
Alar part

Nasalis

Procerus

Nasalis — Transverse part
Alar part

Depressor septi nasi

Procerus

Latin, *procerus*, long.

Origin
Fascia over nasal bone. Upper part of lateral nasal cartilage.

Insertion
Skin between eyebrows.

Nerve supply
Facial nerve (VII) (temporal branch).

Blood supply
Supratrochlear branch of ophthalmic artery
(from internal carotid artery).

Action
Produces transverse wrinkles over bridge of nose. Pulls medial portion of eyebrows downward.

Nasalis

Latin, *nasus*, nose.

Origin
Transverse part
Maxilla just lateral to nose.
Alar part
Maxilla over lateral incisor.

Insertion
Transverse part
Joins muscle of opposite side across bridge of nose.
Alar part
Alar cartilage of nose.

Nerve supply
Facial nerve (VII) (buccal branch).

Blood supply
Superior labial branch of the facial artery
(from external carotid artery).

Action
Transverse part
Compresses nasal aperture.
Alar part
Draws cartilage downward and laterally, opening nostril.

Depressor Septi Nasi

Latin, *deprimere*, to press down; *septem*, seven; *nasi*, of the nose.

Origin
Maxilla above medial incisor.

Insertion
Nasal septum and ala.

Nerve supply
Facial nerve (VII) (buccal branch).

Blood supply
Superior labial branch of the facial artery
(from external carotid artery).

Action
Pulls the nose inferiorly, so assisting nasalis in opening the nostrils.

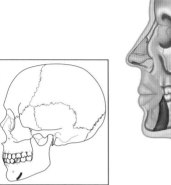

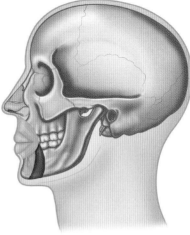

Depressor Anguli Oris

Latin, *deprimere*, to press down; *anguli*, of the corner; *oris*, mouth.

Origin
Oblique line of mandible.

Insertion
Skin at corner of mouth.

Nerve supply
Facial nerve (VII) (mandibular and buccal branches).

Blood supply
Inferior labial and submental branches of the facial artery (from external carotid artery).

Action
Pulls corner of mouth downward and laterally.

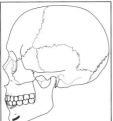

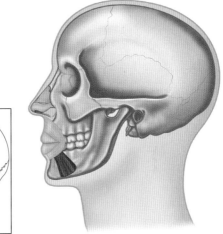

Depressor Labii Inferioris

Latin, *deprimere*, to press down; *labii*, of the lip; *inferioris*, of the lower.

Origin
Anterior part of oblique line of mandible.

Insertion
Skin of lower lip.

Nerve supply
Facial nerve (VII) (mandibular branch).

Blood supply
Inferior labial and submental branches of the facial artery (from external carotid artery).

Action
Pulls lower lip downward and laterally.

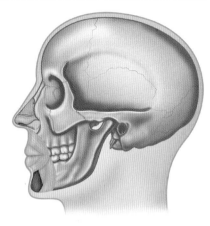

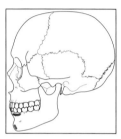

Mentalis

Latin, *mentum*, chin.

Origin
Mandible inferior to incisor teeth.

Insertion
Skin of chin.

Nerve supply
Facial nerve (VII) (mandibular branch).

Blood supply
Inferior labial and submental branches of the facial artery (from external carotid artery).

Action
Protrudes lower lip and pulls up (wrinkles) skin of chin.

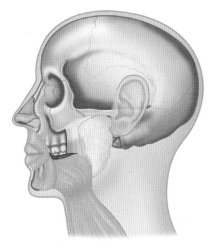

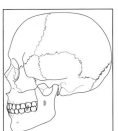

Risorius

Latin, *risus*, laughter.

Origin
Fascia over masseter muscle.

Insertion
Skin at corner of mouth.

Nerve supply
Facial nerve (VII) (buccal branch).

Blood supply
Transverse facial artery and facial artery
(from external carotid artery).

Action
Retracts corner of mouth.

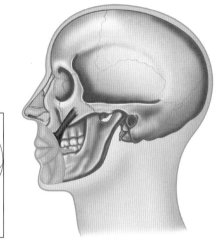

Zygomaticus Major

Greek, *zygoma*, bar, bolt. **Latin**, *major*, larger.

Origin
Posterior part of lateral surface of zygomatic bone.

Insertion
Skin at corner of mouth.

Nerve supply
Facial nerve (VII) (zygomatic and buccal branches).

Blood supply
Transverse facial artery and facial artery
(from external carotid artery).

Action
Pulls corner of mouth upward and laterally, as in smiling.

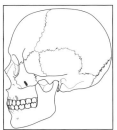

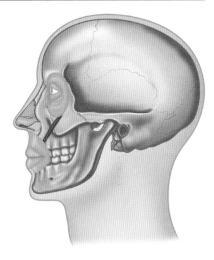

Zygomaticus Minor

Greek, *zygoma*, bar, bolt. **Latin**, *minor*, smaller.

Origin
Anterior part of lateral surface of zygomatic bone.

Insertion
Upper lip just medial to corner of mouth.

Nerve supply
Facial nerve (VII) (buccal branch).

Blood supply
Transverse facial artery and facial artery
(from external carotid artery).

Action
Elevates upper lip.

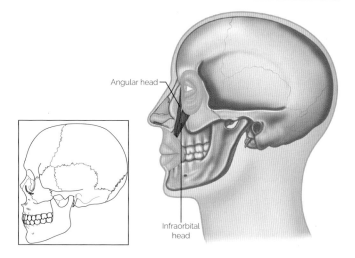

Angular head

Infraorbital head

Levator Labii Superioris

Latin, *levare*, to lift; *labii*, of the lip; *superioris*, of the upper.

Origin
Angular head
Zygomatic bone and frontal process of maxilla.
Infraorbital head
Lower border of orbit.

Insertion
Angular head
Greater alar cartilage, upper lip, and skin of nose.
Infraorbital head
Muscles of upper lip.

Nerve supply
Facial nerve (VII) (buccal branch).

Blood supply
Angular head
Infraorbital artery
(via maxillary artery, from external carotid artery).
Infraorbital head
Superior labial branch of the facial artery
(from external carotid artery).

Action
Raises upper lip. Dilates nostril. Forms nasolabial furrow.

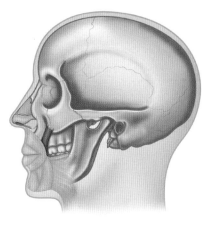

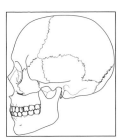

Levator Anguli Oris

Latin, *levare*, to lift; *anguli*, of the corner; *oris*, mouth.

Origin
Canine fossa of maxilla.

Insertion
Skin at corner of mouth.

Nerve supply
Facial nerve (VII) (buccal branch).

Blood supply
Superior labial branch of the facial artery
(from external carotid artery).

Action
Elevates angle (corner) of mouth.
Helps form nasolabial furrow.

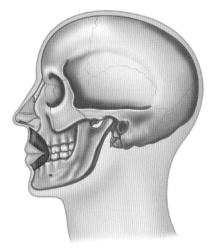

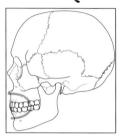

Orbicularis Oris

Latin, *orbiculus*, small circular disc; *oris*, mouth.

Origin
Muscle fibers surrounding opening of mouth, attached to skin, muscle, and fascia of lips and surrounding area.

Insertion
Skin and fascia at corner of mouth.

Nerve supply
Facial nerve (VII) (buccal and mandibular branches).

Blood supply
Superior and inferior labial branches of the facial artery (from external carotid artery).

Action
Closes lips. Compresses lips against teeth. Protrudes (purses) lips. Shapes lips during speech.

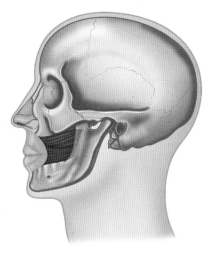

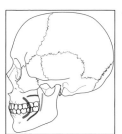

Buccinator

Latin, *bucca*, cheek.

Origin
Posterior parts of maxilla and mandible; pterygomandibular raphe.

Insertion
Blends with orbicularis oris and into lips.

Nerve supply
Facial nerve (VII) (buccal branch).

Blood supply
Facial artery
(from external carotid artery).

Action
Presses cheek against teeth. Compresses distended cheeks.

Muscles of Mastication

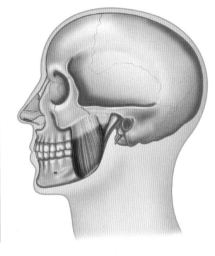

Masseter

Greek, *maseter*, chewer.

Origin
Zygomatic arch and maxillary process of zygomatic bone.

Insertion
Lateral surface of ramus of mandible.

Nerve supply
Trigeminal nerve (V) (mandibular division).

Blood supply
Masseteric branch of the maxillary artery
(from external carotid artery).

Action
Elevation of mandible.

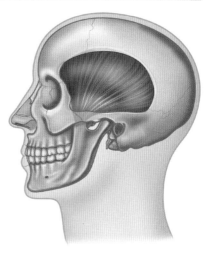

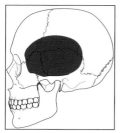

Temporalis

Latin, *temporalis*, of time.

Origin
Bone of temporal fossa. Temporal fascia.

Insertion
Coronoid process of mandible. Anterior margin of ramus of mandible.

Nerve supply
Anterior and posterior deep temporal nerves from the trigeminal nerve (V) (mandibular division).

Blood supply
Anterior and posterior deep temporal branches of maxillary artery
(from external carotid artery).

Action
Elevation and retraction of mandible.

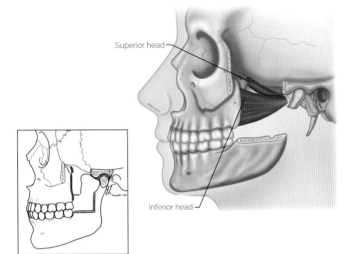

Superior head

Inferior head

Lateral Pterygoid

Greek, *pterygoeides*, wing-like.
Latin, *lateralis*, relating to the
side.

Origin
Superior head
Roof of infratemporal fossa.
Inferior head
Lateral surface of lateral plate
of pterygoid process.

Insertion
Superior head
Capsule and articular disc of
temporomandibular joint.
Inferior head
Neck of mandible.

Nerve supply
Trigeminal nerve (V)
(mandibular division).

Blood supply
**Lateral pterygoid branch of the
maxillary artery**
(from external carotid artery).

Action
Protrusion and side-to-side
movements of mandible, as in
chewing.

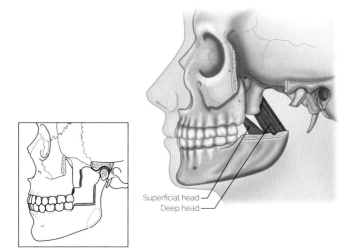

Superficial head
Deep head

Medial Pterygoid

Greek, *pterygoeides*, wing-like.
Latin, *medialis*, relating to the middle.

Origin
Deep head
Medial surface of lateral pterygoid plate of pterygoid process. Pyramidal process of palatine bone.
Superficial head
Tuberosity of maxilla and pyramidal process of palatine bone.

Insertion
Medial surface of ramus and angle of mandible.

Nerve supply
Trigeminal nerve (V)
(mandibular division).

Blood supply
Medial pterygoid branch of the maxillary artery
(from external carotid artery).

Action
Elevation and side-to-side movement of mandible, as in chewing.

CHAPTER 4

Muscles of the Neck

The neck is made up of five sections of tissue running longitudinally:

1. The cervical spine, surrounded by muscles (a musculovertebral block) and enclosed in prevertebral fascia.
2. The pharynx and larynx, enclosed by pretracheal fascia.
3. & 4. Two vascular blocks. These are left- and right-sided fascial sheaths surrounding the common and internal carotid arteries, the internal jugular vein, and the vagus nerve.
5. The outer investing layer of fascia, enclosing the sternocleidomastoid and trapezius muscles.

Sternocleidomastoid (SCM) divides the neck into two regions—the anterior triangle and posterior triangle. The anterior and posterior triangles of the neck are anatomical divisions created by the muscles of the head and neck. It is important to note that all triangles mentioned here are paired—they appear on the left and right sides of the neck.

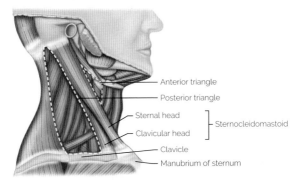

Platysma

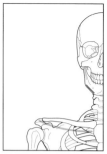

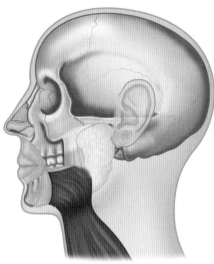

Platysma

Greek, *platys*, broad, flat.

Origin
Subcutaneous fascia of upper quarter of chest.

Insertion
Subcutaneous fascia and muscles of chin and jaw. Inferior border of mandible.

Nerve supply
Facial nerve (VII) (cervical branch).

Blood supply
Facial artery
(from external carotid artery).

Action
Pulls lower lip from corner of mouth downward and laterally. Draws skin of chest upward.

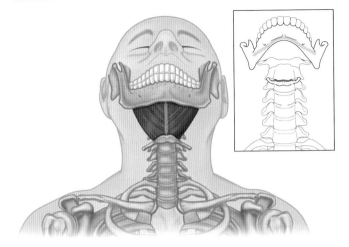

Mylohyoid

Greek, *mylos*, millstone, molar; *hyoeides*, shaped like the Greek letter upsilon (υ).

Origin
Mylohyoid line on inner surface of mandible.

Insertion
Median fibrous raphe and adjacent part of hyoid bone.

Nerve supply
Mylohyoid nerve from inferior alveolar branch of trigeminal V nerve (mandibular division).

Blood supply
Mylohyoid branch of the inferior alveolar branch of the maxillary artery (from external carotid artery).

Action
Depresses mandible when hyoid is fixed. Elevates and pulls hyoid forward when mandible is fixed. Supports and elevates floor of oral cavity.

Anterior Triangle—Suprahyoid Muscles

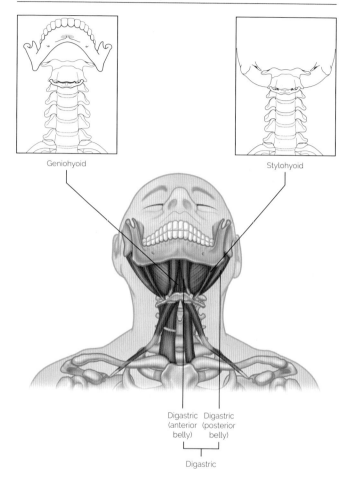

Geniohyoid

Stylohyoid

Digastric (anterior belly) Digastric (posterior belly)

Digastric

Geniohyoid

Greek, *geneion*, chin; *hyoeides*, shaped like the Greek letter upsilon (υ).

Origin
Inferior mental spine on inner surface of mandible.

Insertion
Hyoid bone.

Nerve supply
Branch from ventral ramus of C1 carried along hypoglossal nerve (XII).

Blood supply
Lingual artery, and submental branch of the facial artery (from external carotid artery).

Action
Protrudes and elevates hyoid bone, widening pharynx for reception of food. Depresses mandible if hyoid bone is fixed.

Stylohyoid

Latin, *stilus*, stake, pale. **Greek**, *hyoeides*, shaped like the Greek letter upsilon (υ).

Origin
Base of styloid process of temporal bone.

Insertion
Hyoid bone (after splitting to enclose the intermediate tendon of digastric).

Nerve supply
Facial nerve (VII) (mandibular branches).

Blood supply
Ascending pharyngeal artery, and may receive supply from branches of facial artery (from external carotid artery).

Action
Pulls hyoid bone upward and backward, thereby elevating tongue.

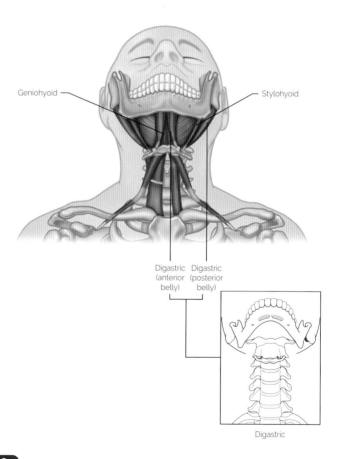

Geniohyoid

Stylohyoid

Digastric (anterior belly)

Digastric (posterior belly)

Digastric

Digastric

Latin, *digastricus*, having two (muscle) bellies.

Origin
Anterior belly
Digastric fossa on inner side of lower border of mandible.
Posterior belly
Mastoid notch on medial side of mastoid process of temporal bone.

Insertion
Body of hyoid bone via a fascial sling over an intermediate tendon.

Nerve supply
Anterior belly
Mylohyoid nerve, from trigeminal V nerve (mandibular division).

Posterior belly
Facial nerve (VII) (digastric branch).

Blood supply
Auricular, occipital, and stylomastoid branches of the posterior auricular artery (from external carotid artery).

Action
Anterior belly
Raises hyoid bone. Opens mouth by lowering mandible.
Posterior belly
Pulls hyoid upward and back.

Anterior Triangle—Infrahyoid Muscles

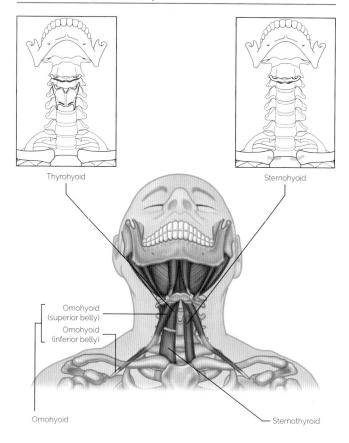

Thyrohyoid

Sternohyoid

Omohyoid
(superior belly)

Omohyoid
(inferior belly)

Omohyoid

Sternothyroid

Thyrohyoid

Greek, *thyreos*, oblong shield; *hyoeides*, shaped like the Greek letter upsilon (υ).

Origin
Oblique line of outer surface of thyroid cartilage.

Insertion
Lower border of body and greater horn of hyoid bone.

Nerve supply
Fibers from ventral ramus of C1 carried along hypoglossal nerve (XII).

Blood supply
Superior thyroid artery (from external carotid artery). Can also receive supply from inferior thyroid artery (from thyrocervical trunk of the subclavian artery).

Action
Raises thyroid and depresses hyoid bone, thus closing laryngeal orifice, preventing food from entering larynx during swallowing.

Sternohyoid

Greek, *sternon*, chest; *hyoeides*, shaped like the Greek letter upsilon (υ).

Origin
Posterior aspect of sternoclavicular joint, and adjacent manubrium of sternum.

Insertion
Lower border of hyoid bone (medial to insertion of omohyoid).

Nerve supply
Ventral rami of C1 to 3 through the ansa cervicalis.

Blood supply
Superior thyroid artery (from external carotid artery). Can also receive supply from inferior thyroid artery (from thyrocervical trunk of the subclavian artery).

Action
Depresses hyoid bone after swallowing.

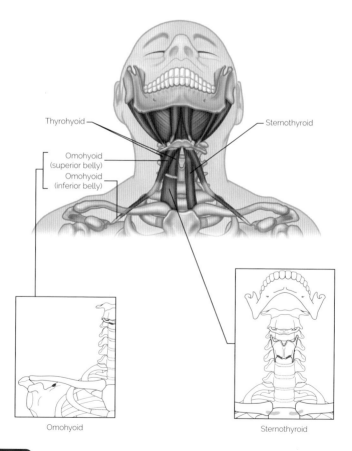

Thyrohyoid

Sternothyroid

Omohyoid
(superior belly)

Omohyoid
(inferior belly)

Omohyoid

Sternothyroid

Omohyoid

Greek, *omos*, shoulder; *hyoeides*, shaped like the Greek letter upsilon (υ).

Origin
Inferior belly
Upper border of scapula medial to the scapular notch.
Superior belly
Intermediate tendon.

Insertion
Inferior belly
Intermediate tendon.
Superior belly
Lower border of hyoid bone, lateral to insertion of sternohyoid.

Note: The intermediate tendon is tied down to the clavicle and first rib by a sling of the cervical fascia.

Nerve supply
Ventral rami of C1 to C3 through ansa cervicalis.

Blood supply
Transverse cervical artery
(from subclavian artery).
Can also receive supply from inferior thyroid artery (from thyrocervical trunk of the subclavian artery).

Action
Depresses and fixes hyoid bone.

Sternothyroid

Greek, *sternon*, chest; *thyreos*, oblong shield.

Origin
Posterior surface of manubrium of sternum.

Insertion
Oblique line on outer surface of thyroid cartilage.

Nerve supply
Ventral rami of C1 to 3 through the ansa cervicalis.

Blood supply
Superior thyroid artery
(from external carotid artery).
Can also receive supply from inferior thyroid artery (from thyrocervical trunk of the subclavian artery).

Action
Draws larynx downward.

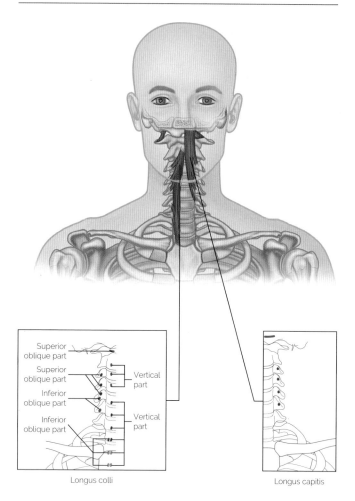

Superior oblique part
Superior oblique part
Inferior oblique part
Inferior oblique part

Vertical part

Vertical part

Longus colli

Longus capitis

Longus Colli

Latin, *longus*, long; *colli*, of the neck.

Origin
Superior oblique
Transverse processes of C3–5.
Inferior oblique
Anterior surface of bodies of T1, 2, maybe T3.
Vertical
Anterior surface of bodies of T1–3 and C5–7.

Insertion
Superior oblique
Anterior arch of atlas.
Inferior oblique
Transverse processes of C5–6.
Vertical
Transverse processes of C2–4.

Nerve supply
Ventral rami of cervical nerves C2–6.

Blood supply
Deep cervical artery of costocervical trunk
(from subclavian artery).

Action
Flexes neck anteriorly and laterally and slight rotation to opposite side.

Longus Capitis

Latin, *longus*, long; *capitis*, of the head.

Origin
Transverse processes of C3–6.

Insertion
Inferior surface of basilar part of occipital bone.

Nerve supply
Ventral rami of cervical nerves C1–3, (C4).

Blood supply
Deep cervical artery of costocervical trunk
(from subclavian artery).

Action
Flexes head.

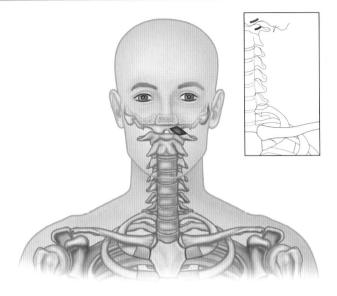

Rectus Capitis Anterior

Latin, *rectus*, straight; *capitis*, of the head; *anterior*, at the front.

Origin
Anterior surface of lateral mass of atlas and its transverse process.

Insertion
Inferior surface of basilar part of occipital bone.

Nerve supply
Branches from ventral rami of cervical nerves C1, 2.

Blood supply
Deep cervical artery of costocervical trunk (from subclavian artery).

Action
Flexes head at atlanto-occipital joint.

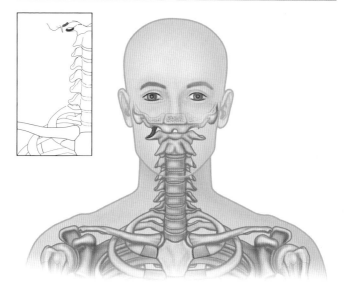

Rectus Capitis Lateralis

Latin, *rectus*, straight; *capitis*, of the head; *lateralis*, relating to the side.

Origin
Transverse process of atlas.

Insertion
Jugular process of occipital bone.

Nerve supply
Branches from ventral rami of cervical nerves C1, 2.

Blood supply
Deep cervical artery of costocervical trunk (from subclavian artery).

Action
Flexes head laterally to same side. Stabilizes atlanto-occipital joint.

Posterior Triangle

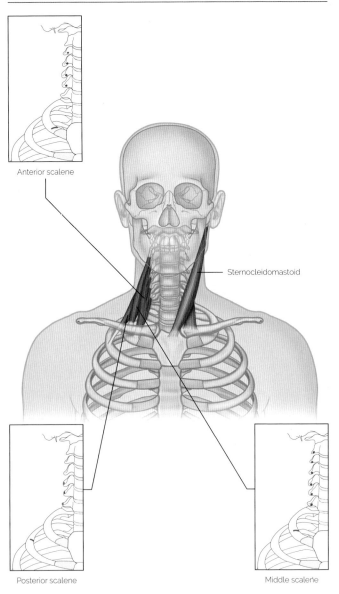

Anterior scalene

Sternocleidomastoid

Posterior scalene

Middle scalene

Scalenes

Greek, *skalenos*, uneven. **Latin**, *anterior*, at the front; *medius*, middle; *posterior*, at the back.

Origin
Anterior
Anterior tubercles of transverse processes of C3–6.
Middle
Transverse processes of C2–7.
Posterior
Posterior tubercles of transverse processes of C4–6.

Insertion
Anterior
Scalene tubercle and upper surface of first rib.
Middle
Upper surface of first rib, behind groove for subclavian artery.
Posterior
Upper surface of second rib.

Nerve supply
Anterior
Ventral rami of cervical nerves C4–7.
Middle
Ventral rami of cervical nerves C3–7.
Posterior
Ventral rami of lower cervical nerves C5–7.

Blood supply
Anterior
Inferior thyroid artery of the thyrocervical trunk (from subclavian artery).

Middle and posterior
Ascending cervical branch of the inferior thyroid artery (from thyrocervical trunk of subclavian artery).

Action
Acting on both sides
Flex neck; raise first or second rib during active respiratory inhalation.
Acting on one side
Side flex and rotate head.

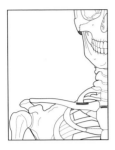

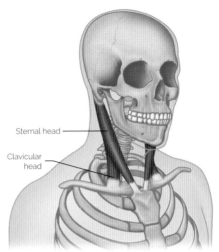

Sternal head

Clavicular head

Sternocleidomastoid

Greek, *sternon*, chest; *kleis*, key; *mastoeides*, breast shaped.

Origin
Sternal head
Upper part of anterior surface of manubrium of sternum.
Clavicular head
Upper surface of medial third of clavicle.

Insertion
Sternal head
Lateral one-half of superior nuchal line of occipital bone.
Clavicular head
Outer surface of mastoid process of temporal bone.

Nerve supply
Accessory nerve (XI) and branches from ventral rami of cervical nerves C2, 3 (C4).

Blood supply
Sternocleidomastoid branches of the occipital artery and superior thyroid arteries (from external carotid artery).

Action
Bilateral contraction
Draws head forward (protracts); raises sternum, and consequently ribs, during deep inhalation.
Unilateral contraction
Flexes head to same side; rotates head to opposite side.

CHAPTER 5

Muscles of the Trunk

Muscles of the Back

The muscles of the back can be divided into:

- Superficial—associated with movements of the shoulder.
- Intermediate—associated with movements of the thoracic cage and respiration.
- Deep—associated with movements of the vertebral column.

The superficial and intermediate muscles do not develop in the back and are classified as *extrinsic muscles*; they are involved in moving the upper limbs and thoracic wall. The superficial extrinsic muscles form the V-shaped musculature associated with the middle and upper back, and include the trapezius, latissimus dorsi, levator scapulae, and rhomboids. The intermediate extrinsic muscles run from the vertebral column to the ribcage and assist with elevating and depressing the ribs. They are thought to have a slight respiratory function.

The deep muscles develop embryologically in the back and are thus described as *intrinsic muscles*. They are involved with maintaining posture and allow the upper body and vertebral column to move in flexion, lateral flexion, extension, hyperextension, and rotation. This deep intrinsic group of muscles can be further subdivided into superficial, intermediate, and deep layers.

The superficial layer is located on the posterolateral portions of the neck covering deeper muscles, and they laterally flex, rotate, and extend the head and neck.

The **erector spinae** muscle forms the intermediate layer of the deep intrinsic muscles. Underneath the erector spinae muscles is another layer of muscles that help to support posture and assist the intermediate muscles in moving the spine. The deep intrinsic muscles are a group of short muscles associated with the transverse and spinous processes of the vertebral column; none of these muscles traverses more than six vertebral segments.

Postvertebral Muscles—Erector Spinae—Iliocostalis Portion

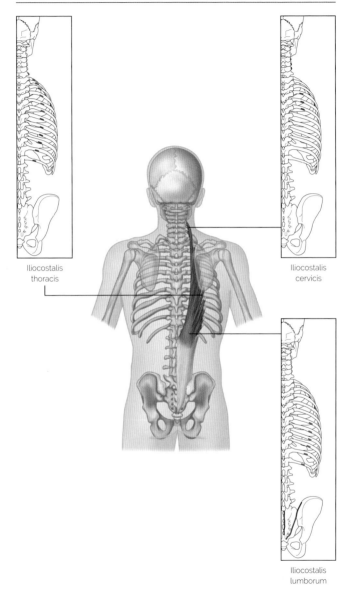

Iliocostalis
thoracis

Iliocostalis
cervicis

Iliocostalis
lumborum

Latin, *iliocostalis*, from ilium to rib; *lumborum*, of the loins; *thoracis*, of the chest; *cervicis*, of the neck.

Origin

Lumborum
Sacrum, spinous processes of lumbar and lower two thoracic vertebrae, and their supraspinous ligaments, and the iliac crest.
Thoracis
Angles of lower six ribs, medial to iliocostalis lumborum.
Cervicis
Angles of third to sixth ribs.

Insertion

Lumborum
Angles of lower six or seven ribs.
Thoracis
Angles of upper six ribs and transverse process of C7.
Cervicis
Transverse processes of C4–6.

Nerve supply

Dorsal rami of cervical, thoracic and lumbar spinal nerves.

Blood supply

Lumborum
Lumbar arteries
(from abdominal aorta).
Subcostal arteries
(from thoracic aorta).
Thoracis
Posterior intercostal and subcostal arteries
(from thoracic aorta).
Cervicis
Deep cervical artery of costocervical trunk
(from subclavian artery).

Action

Extends and side flexes vertebral column. Draws ribs down for forceful inhalation (thoracis only).

Postvertebral Muscles—Erector Spinae—Longissimus Portion

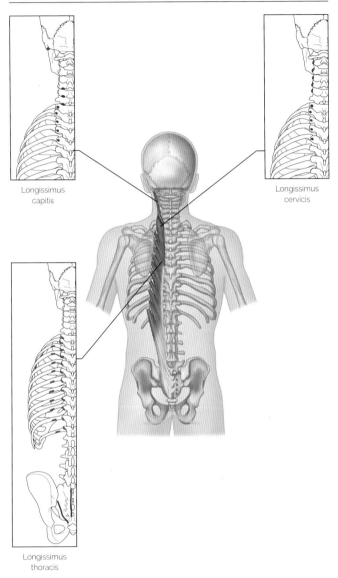

Longissimus
capitis

Longissimus
cervicis

Longissimus
thoracis

Latin, *longissimus*, longest; *thoracis*, of the chest; *cervicis*, of the neck; *capitis*, of the head.

Origin

Thoracis
Blends with iliocostalis in lumbar region and is attached to transverse processes of lumbar vertebrae.
Cervicis
Transverse processes of T1–5.
Capitis
Transverse processes of T1–5. Articular processes of C4–7.

Insertion

Thoracis
Transverse processes of T1–12. Area between tubercles and angles of lower nine or ten ribs.
Cervicis
Transverse processes of C2–6.
Capitis
Posterior margin of mastoid process of temporal bone.

Nerve supply
Dorsal rami of spinal nerves C1–S1.

Blood supply

Thoracis
Supplied segmentally by posterior intercostal and subcostal arteries
(from thoracic aorta).
Lumbar arteries
(from abdominal aorta).
Cervicis and Capitis
Supplied segmentally by deep cervical artery of costocervical trunk
(from subclavian artery).
Posterior intercostal and subcostal arteries
(from thoracic aorta).

Action

Extends and side flexes vertebral column. Draws ribs down for forceful inhalation (thoracic only). Extends and rotates head (capitis only).

Postvertebral Muscles—Erector Spinae—Spinalis Portion

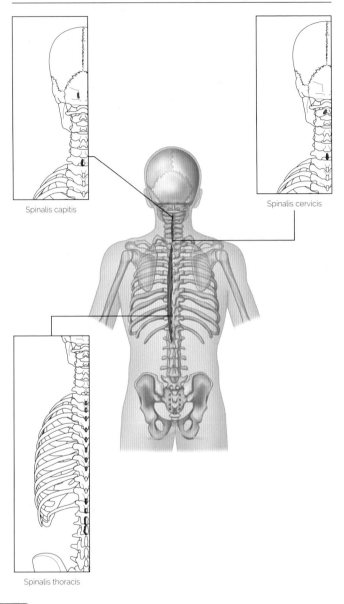

Spinalis capitis

Spinalis cervicis

Spinalis thoracis

Latin, *spinalis*, relating to the spine; *thoracis*, of the chest; *cervicis*, of the neck; *capitis*, of the head.

Origin
Thoracis
Spinous processes of T11–12 and L1–2.
Cervicis
Ligamentum nuchae. Spinous process of C7.
Capitis
Usually blends with semispinalis capitis.

Insertion
Thoracis
Spinous processes of T1–8.
Cervicis
Spinous process of C2.
Capitis
With semispinalis capitis.

Nerve supply
Dorsal rami of spinal nerves C2–L3.

Blood supply
Thoracis
Supplied segmentally by posterior intercostal and subcostal arteries
(from thoracic aorta).
Lumbar arteries
(from abdominal aorta).
Cervicis
Supplied segmentally by deep cervical artery of costocervical trunk
(from subclavian artery).

Action
Extends vertebral column. Helps maintain correct curvature of spine in standing and sitting positions. Extends head (capitis only).

Postvertebral Muscles—Spinotransversales Group

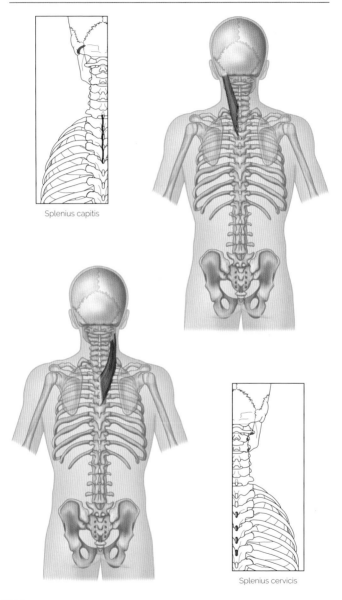

Splenius capitis

Splenius cervicis

Splenius Capitis and Splenius Cervicis

Greek, *splenion*, bandage. **Latin**, *capitis*, of the head; *cervicis*, of the neck.

Origin
Capitis
Lower part of ligamentum nuchae. Spinous processes of C7 and T1–4.
Cervicis
Spinous processes of T3–6.

Insertion
Capitis
Posterior aspect of mastoid process of temporal bone. Lateral part of superior nuchal line, deep to attachment of sternocleidomastoid.
Cervicis
Posterior tubercles of transverse processes of C1–3.

Nerve supply
Capitis
Dorsal rami of middle cervical nerves.
Cervicis
Dorsal rami of lower cervical nerves.

Blood supply
Supplied segmentally by deep cervical artery of costocervical trunk (from subclavian artery).
Posterior intercostal and subcostal arteries (from thoracic aorta).

Action
Acting on both sides
Extend head and neck.
Acting on one side
Side flex neck; rotate head to same side as contracting muscle.

Postvertebral Muscles—Transversospinales Group

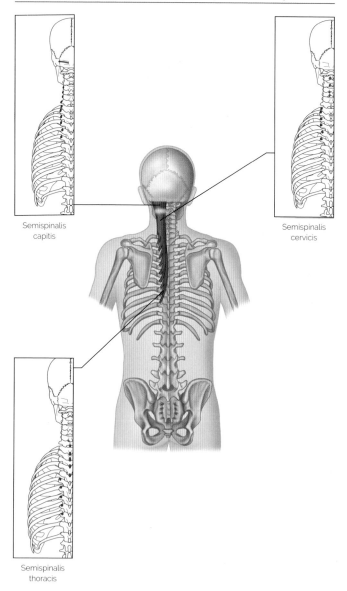

Semispinalis
capitis

Semispinalis
cervicis

Semispinalis
thoracis

Semispinalis

Latin, *semispinalis*, half-spinal; *thoracis*, of the chest; *cervicis*, of the neck; *capitis*, of the head.

Origin
Thoracis
Transverse processes of T6–10.
Cervicis
Transverse processes of T1–6.
Capitis
Transverse processes of C4–T7.

Insertion
Thoracis
Spinous processes of C6–T4.
Cervicis
Spinous processes of C2–5.
Capitis
Between superior and inferior nuchal lines of occipital bone.

Nerve supply
Dorsal rami of thoracic and cervical spinal nerves.

Blood supply
Supplied segmentally by deep cervical artery of costocervical trunk (from subclavian artery).
Posterior intercostal and subcostal arteries (from thoracic aorta).

Action
Extends thoracic and cervical parts of vertebral column. Assists in rotation of thoracic and cervical vertebrae. Semispinalis capitis extends and assists in rotation of the head.

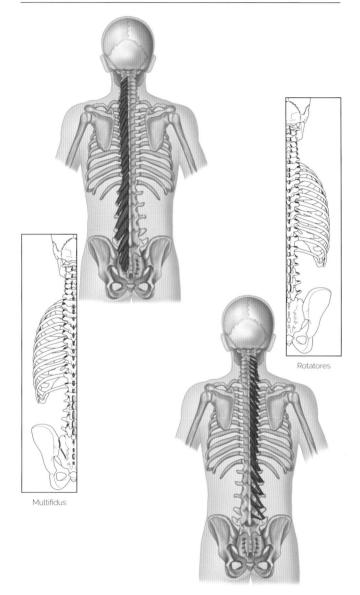

Rotatores

Multifidus

Multifidus

Latin, *multi*, many; *findere*, to split.

Origin
Sacrum, origin of erector spinae, posterior superior iliac spine, mammillary processes (posterior borders of superior articular processes) of L1–5.
Transverse processes of T1–12.
Articular processes of C4–7.

Insertion
Base of spinous processes of L5–C2.

Nerve supply
Dorsal rami of spinal nerves.

Blood supply
Supplied segmentally by deep cervical artery of costocervical trunk
(from subclavian artery).
Posterior intercostal and subcostal arteries
(from thoracic aorta).
Lumbar arteries
(from abdominal aorta).

Action
Gives individual vertebral joints control during movement by the more powerful superficial prime movers. Extension, side flexion, and rotation of vertebral column.

Rotatores

Latin, *rota*, wheel.

Origin
Transverse process of each vertebra.

Insertion
Base of spinous process of adjoining vertebra above.

Nerve supply
Dorsal rami of spinal nerves.

Blood supply
Supplied segmentally by deep cervical artery of costocervical trunk
(from subclavian artery).
Posterior intercostal and subcostal arteries
(from thoracic aorta).

Action
Rotate and assist in extension of vertebral column.

Postvertebral Muscles—Segmental Group

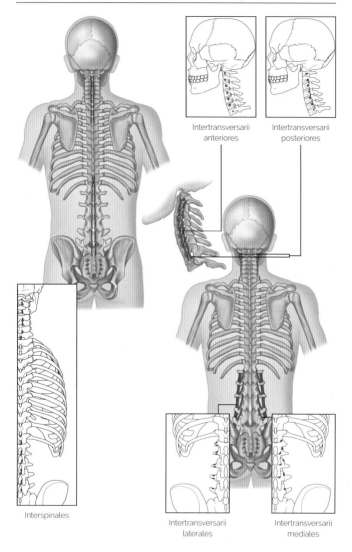

Intertransversarii
anteriores

Intertransversarii
posteriores

Interspinales

Intertransversarii
laterales

Intertransversarii
mediales

Interspinales

Latin, *inter*, between; *spinalis*, relating to the spine.

Origin/Insertion
Extend from one spinous process (origin) to the next one above (insertion) throughout the vertebral column. Positioned either side of interspinous ligament.

Nerve supply
Dorsal rami of spinal nerves.

Blood supply
Supplied segmentally by deep cervical artery of costocervical trunk
(from subclavian artery).
Posterior intercostal and subcostal arteries
(from thoracic aorta).

Action
Postural muscles that stabilize adjoining vertebrae during movements of the vertebral column.

Intertransversarii

Latin, *inter*, between; *transversus*, across, crosswise; *anterior*, at the front; *posterior*, at the back; *lateralis*, relating to the side; *medialis*, relating to the middle.

Origin
Anteriores
Anterior tubercle of transverse processes of T1–C2.
Posteriores
Posterior tubercle of transverse processes of T1–C2.
Laterales
Transverse processes of lumbar vertebrae.
Mediales
Mammillary process (posterior border of superior articular process of lumbar vertebrae).

Insertion
Anteriores
Anterior tubercle of adjacent vertebra above.
Posteriores and Laterales
Transverse process of adjacent vertebra above.
Mediales
Accessory process of adjacent lumbar vertebra above.

Nerve supply
Ventral rami of spinal nerves (apart from mediales, dorsal rami of spinal nerves).

Blood supply
Anteriores and Posteriores
Supplied segmentally by deep cervical artery of costocervical trunk
(from subclavian artery).
Posteriores only
Posterior intercostal and subcostal arteries
(from thoracic aorta).
Laterales and Mediales
Supplied segmentally by lumbar arteries
(from abdominal aorta).

Action
Postural muscles that stabilize adjoining vertebrae during movements of the vertebral column.

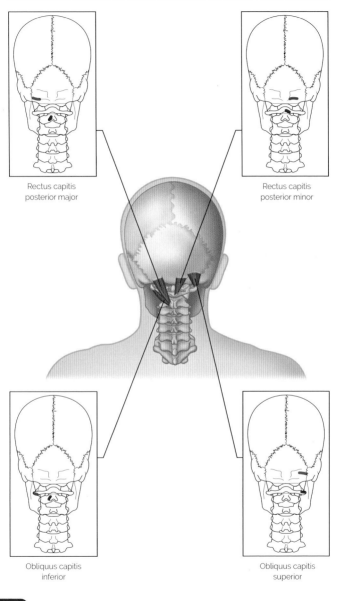

Rectus capitis
posterior major

Rectus capitis
posterior minor

Obliquus capitis
inferior

Obliquus capitis
superior

Rectus Capitis Posterior Major

Latin, *rectus*, straight; *capitis*, of the head; *posterior*, at the back; *major*, larger.

Origin
Spinous process of axis.

Insertion
Lateral portion of occipital bone below inferior nuchal line.

Nerve supply
Suboccipital nerve (dorsal ramus of C1).

Action
Extends head. Rotates head to same side.

Rectus Capitis Posterior Minor

Latin, *minor*, smaller.

Origin
Posterior tubercle of atlas.

Insertion
Medial portion of occipital bone below inferior nuchal line.

Nerve supply
Suboccipital nerve (dorsal ramus of C1).

Action
Extends head.

Obliquus Capitis Inferior

Latin, *obliquus*, diagonal, slanted; *capitis*, of the head; *inferior*, lower.

Origin
Spinous process of axis.

Insertion
Transverse process of atlas.

Nerve supply
Suboccipital nerve (dorsal ramus of C1).

Action
Rotates atlas upon axis, thereby rotating head to same side.

Obliquus Capitis Superior

Latin, *superior*, upper.

Origin
Transverse process of atlas.

Insertion
Occipital bone between superior and inferior nuchal lines.

Nerve supply
Suboccipital nerve (dorsal ramus of C1).

Action
Extends head and flexes to the same side.

Blood supply for suboccipital group
Occipital artery
(from external carotid artery).
Muscular branches of the vertebral artery
(from subclavian artery).

Muscles of the Thorax

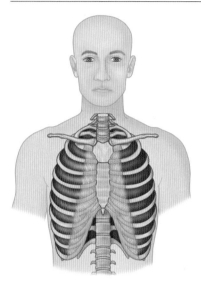

External intercostals

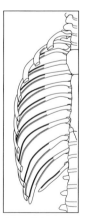

Internal intercostals

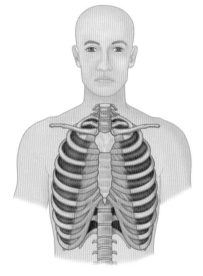

The muscles of the thorax include the diaphragm, which separates the thoracic cavity from the abdominal cavity, as well as the five muscles of the thoracic cage. These muscles are primarily responsible for changing the volume of the thoracic cavity during respiration.

External Intercostals

Latin, *inter*, between; *costa*, rib; *externi*, external.

Origin
Lower border of a rib.

Insertion
Upper border of rib below (fibers run obliquely forward and downward).

Nerve supply
The corresponding intercostal nerves.

Blood supply
Intercostal arteries
(from costocervical trunk of subclavian artery and thoracic aorta).

Action
Contract to stabilize ribcage during various movements of trunk. May elevate ribs during inspiration, thus increasing volume of thoracic cavity (although this action is disputed). Prevent intercostal space from bulging out or sucking in during respiration.

Internal Intercostals

Latin, *inter*, between; *costalis*, relating to the ribs; *interni*, internal.

Origin
Upper border of a rib and costal cartilage.

Insertion
Lower border of rib above (fibers run obliquely forward and upward, toward the costal cartilage).

Nerve supply
The corresponding intercostal nerves.

Blood supply
Intercostal arteries
(from costocervical trunk of subclavian artery and thoracic aorta).

Action
Contract to stabilize ribcage during various movements of trunk. May draw adjacent ribs together during forced expiration, thus decreasing volume of thoracic cavity (although this action is disputed). Prevent intercostal space from bulging out or sucking in during respiration.

Muscles of the Thorax

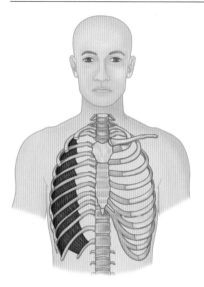

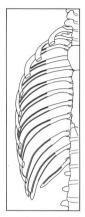

Innermost intercostals

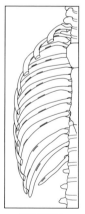

Subcostales

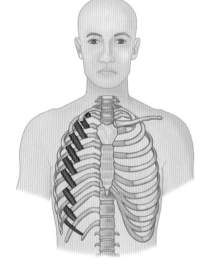

Innermost Intercostals

Latin, *inter*, between; *costalis*, relating to the ribs; *intimo*, innermost part.

Origin
Superior border of each rib.

Insertion
Inferior border of the preceding rib.

Nerve supply
Corresponding intercostal nerves.

Blood supply
Intercostal, internal thoracic, and musculophrenic arteries (from costocervical trunk of subclavian artery and thoracic aorta).

Action
While the action of the innermost intercostals is unknown, it is accepted that they act to fix the position of the ribs during respiration.

Subcostales

Latin, *sub*, under; *costalis*, relating to the ribs.

Origin
Inner surface of each lower rib near its angle.

Insertion
Fibers run obliquely and medially into the inner surface of second or third rib below.

Nerve supply
Corresponding intercostal nerves.

Blood supply
Intercostal arteries (from costocervical trunk of subclavian artery and thoracic aorta).

Action
Contract to stabilize ribcage during various movements of trunk. May draw adjacent ribs together during forced expiration, thus decreasing volume of thoracic cavity (although this action is disputed).

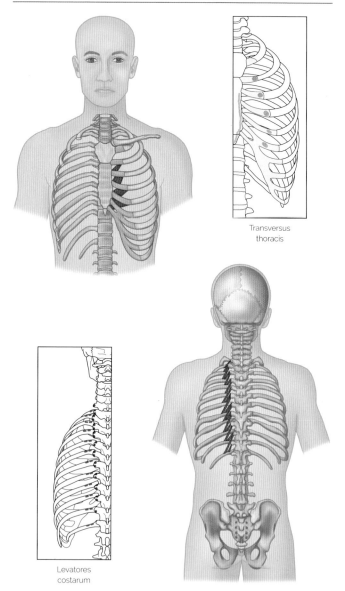

Transversus thoracis

Levatores costarum

Transversus Thoracis

Latin, *transversus*, across, crosswise; *thoracis*, of the chest.

Origin
Posterior surface of xiphoid process and body of sternum. Costal cartilages of fourth to seventh ribs.

Insertion
Inner surfaces of costal cartilages of second to sixth ribs.

Nerve supply
The corresponding intercostal nerves.

Blood supply
Internal thoracic artery
(from subclavian artery).

Action
Draws costal cartilages downward, contributing to forceful exhalation.

Levatores Costarum

Latin, *levare*, to lift; *costarum*, of the ribs.

Origin
Transverse processes of C7–T11.

Insertion
Laterally downward to external surface of rib below, between tubercle and angle.

Nerve supply
Ventral rami of thoracic spinal nerves.

Blood supply
Deep cervical artery of costocervical trunk
(from subclavian artery).
Intercostal arteries
(from costocervical trunk of subclavian artery and thoracic aorta).

Action
Raise the ribs. May very slightly assist side flexion and rotation of vertebral column.

Muscles of the Thorax

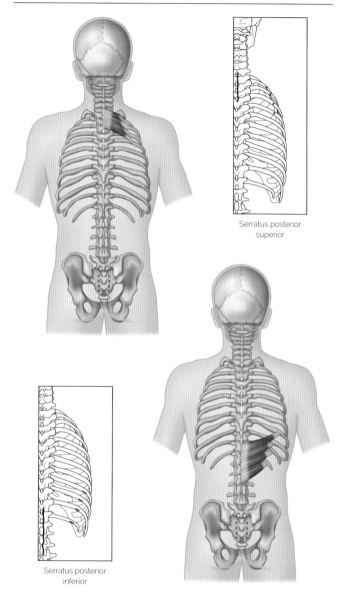

Serratus posterior superior

Serratus posterior inferior

Serratus Posterior Superior

Latin, *serratus*, serrated; *posterior*, at the back; *superior*, upper.

Origin
Lower part of ligamentum nuchae. Spinous processes of C7, T1–3. Supraspinous ligaments.

Insertion
Upper borders of second to fifth ribs, lateral to their angles.

Nerve supply
Ventral rami of upper thoracic nerves T2–5.

Blood supply
Intercostal arteries
(from costocervical trunk of subclavian artery and thoracic aorta).

Action
Raises upper ribs (probably during forced inhalation).

Serratus Posterior Inferior

Latin, *serratus*, serrated; *posterior*, at the back; *inferior*, lower.

Origin
Thoracolumbar fascia, at its attachment to spinous processes of T11–12 and L1–3.

Insertion
Lower borders of last four ribs.

Nerve supply
Ventral rami of lower thoracic nerves T9–12.

Blood supply
Intercostal arteries
(from thoracic aorta).

Action
May help draw lower ribs downward and backward, resisting the pull of the diaphragm.

Muscles of the Thorax

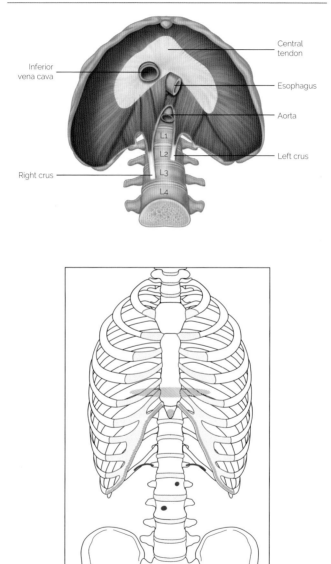

Central tendon

Inferior vena cava

Esophagus

Aorta

Left crus

Right crus

L1

L2

L3

L4

Diaphragm

Greek, *dia*, across; *phragma*, partition, wall.

Origin
Sternal portion
Back of xiphoid process.
Costal portion
Inner surfaces of lower six ribs and their costal cartilages.
Lumbar portion
L1–3. Medial and lateral arcuate ligaments.

Insertion
All fibers converge and attach onto a central tendon.

Nerve supply
Phrenic nerve (ventral rami) C3–5.

Blood supply
Musculophrenic artery
via internal thoracic artery (from subclavian artery).
Superior phrenic artery
(from thoracic aorta).
Inferior phrenic artery
(from abdominal aorta).

Action
Forms floor of thoracic cavity. Pulls central tendon downward during inhalation, thereby increasing volume of thoracic cavity.

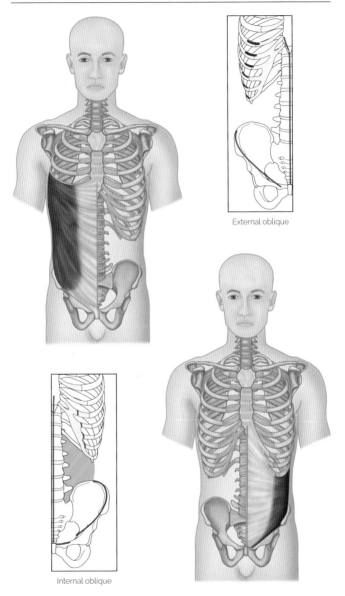

External oblique

Internal oblique

Anteriorly the abdominal wall is formed above by the lower part of the thoracic cage and below by muscle layers.

Posteriorly, the abdominal wall is made up of five lumbar vertebrae and their intervertebral discs; more laterally, it comprises the twelfth ribs and the upper part of the pelvis. Deep to this bony layer is a muscular layer made up of quadratus lumborum, and the two psoas major muscles.

External and Internal Obliques

Latin, *obliquus*, diagonal, slanted; *externus*, external; *internus*, internal; *abdominis*, of the belly/stomach.

Origin
External
Muscular slips from the outer surfaces of the lower eight ribs.
Internal
Iliac crest. Lateral two-thirds of inguinal ligament. Thoracolumbar fascia.

Insertion
External
Lateral lip of iliac crest. Aponeurosis ending in linea alba.
Internal
Inferior borders of bottom three or four ribs. Linea alba via an abdominal aponeurosis. Pubic crest and pectineal line.

Nerve supply
External
Ventral rami of thoracic spinal nerves T5–12.
Internal
Ventral rami of thoracic spinal nerves T7–12 and L1.

Blood supply
Musculophrenic and superior epigastric arteries
(via internal thoracic artery from subclavian artery).
Intercostal arteries 7–11 and subcostal artery
(from thoracic aorta).
Lumbar arteries
(from abdominal aorta).
Superficial circumflex, superficial epigastric, and superficial external pudendal arteries
(from femoral artery).
Deep circumflex iliac and inferior epigastric arteries
(from external iliac artery).

Action
External
Compresses abdomen, helping to support abdominal viscera against pull of gravity. Contraction of one side alone side flexes trunk to that side and rotates it to the opposite side.
Internal
Compresses abdomen, helping to support abdominal viscera against pull of gravity. Contraction of one side alone side flexes and rotates trunk.

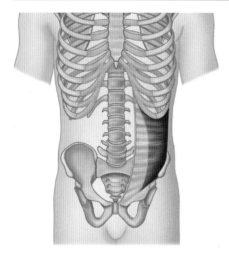 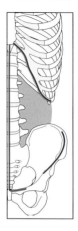

Transversus Abdominis

Latin, *transversus*, across, crosswise; *abdominis*, of the belly/stomach.

Origin
Anterior two-thirds of iliac crest. Lateral third of inguinal ligament. Thoracolumbar fascia. Costal cartilages of lower six ribs.

Insertion
Aponeurosis ending in linea alba. Pubic crest and pectineal line.

Nerve supply
Ventral rami of thoracic spinal nerves T7–12 and L1.

Blood supply
Musculophrenic and superior epigastric arteries via internal thoracic artery (from subclavian artery).
Intercostal arteries 7–11 and subcostal artery (from thoracic aorta).
Lumbar arteries (from abdominal aorta).
Superficial circumflex, superficial epigastric, and superficial external pudendal arteries (from femoral artery).
Deep circumflex iliac and inferior epigastric arteries (from external iliac artery).

Action
Compresses abdomen, helping to support abdominal viscera against pull of gravity.

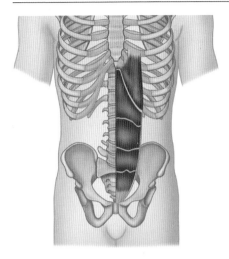
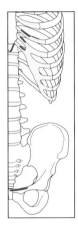

Rectus Abdominis

Latin, *rectus*, straight; *abdominis*, of the belly/stomach.

Origin
Pubic crest, pubic tubercle, and symphysis pubis.

Insertion
Anterior surface of xiphoid process. Fifth, sixth, and seventh costal cartilages.

Nerve supply
Ventral rami of thoracic nerves T5–12.

Blood supply
Superior epigastric artery via internal thoracic artery (from subclavian artery).
Intercostal and subcostal arteries (from thoracic aorta).
Inferior epigastric artery (from external iliac artery).

Action
Flexes lumbar spine and pulls ribcage down. Stabilizes pelvis during walking.

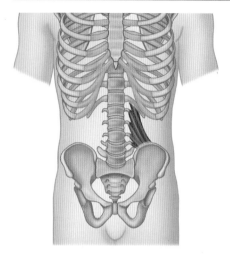

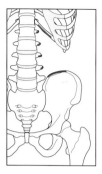

Quadratus Lumborum

Latin, *quadratus*, squared; *lumborum*, of the lower back.

Origin
Transverse process of L5. Posterior part of iliac crest. Iliolumbar ligament.

Insertion
Medial part of lower border of twelfth rib. Transverse processes of L1–4.

Nerve supply
Ventral rami of T12, L1–4.

Blood supply
Subcostal artery (from thoracic aorta).
Lumbar arteries (from abdominal aorta).

Action
Side flexes vertebral column. Fixes twelfth rib during deep respiration (e.g., helps stabilize diaphragm for singers exercising voice control). Helps extend lumbar part of vertebral column and gives it lateral stability.

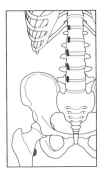

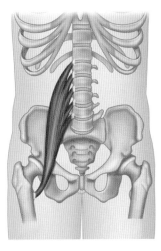

Psoas Major

Greek, *psoa*, muscle of the loins.
Latin, *major*, larger.

Origin
Transverse processes of L1–5.
Bodies of T12–L5 and the
intervertebral discs between each
vertebra.

Insertion
Lesser trochanter of femur.

Nerve supply
Ventral rami of lumbar nerves
L1–3 (psoas minor innervated
from L1, 2).

Blood supply
Subcostal artery
(from thoracic aorta).
Lumbar arteries
(from abdominal aorta).

Action
Main flexors of hip joint.
Flex and laterally rotate thigh,
as in kicking a football. Bring leg
forward in walking or running.
Acting from their insertion, they
flex the trunk, as in sitting up
from supine position.

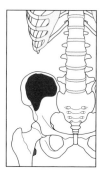

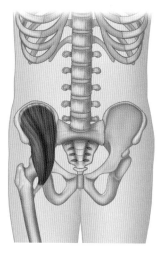

Iliacus

Latin, *iliacus*, relating to the loin.

Origin
Superior two-thirds of iliac fossa. Anterior sacroiliac and iliolumbar ligaments. Upper lateral part of sacrum.

Insertion
Lesser trochanter of femur.

Nerve supply
Femoral nerve L2–4.

Blood supply
Iliolumbar branch of the internal iliac artery
Via common iliac artery (from abdominal aorta).

Action
Main flexors of hip joint. Flex and laterally rotate thigh, as in kicking a football. Bring leg forward in walking or running. Acting from their insertion, they flex the trunk, as in sitting up from supine position.

CHAPTER 6

Muscles of the Shoulder and Arm

Muscles of the Shoulder

Muscles of the shoulder originate in both the back and the anterior thorax and are as follows:

- Muscles of the back that move the shoulder – **trapezius, levator scapulae, rhomboid major, rhomboid minor, latissimus dorsi**, and the **rotator cuff** muscles (see below).
- Muscles of the anterior thorax that move the shoulder – **subclavius, pectoralis minor**, and **serratus anterior**.

Four muscles work together to hold the head of the humerus in the glenoid cavity of the scapula. These muscles are the **subscapularis, supraspinatus, infraspinatus**, and **teres minor**. Together, their tendons form the *rotator cuff*, which encircles the ball-and-socket joint of the shoulder and strengthens and reinforces it.

Muscles of the Arm

The arm is divided by a fascial layer known as the medial and lateral intermuscular septa; this layer divides the arm into anterior and posterior compartments.

Brachialis and **biceps brachii** lie anteriorly and are the major flexors at the elbow, along with **coracobrachialis**. The three-headed **triceps brachii** muscle lies posteriorly and is the major extensor of the elbow, together with **anconeus**.

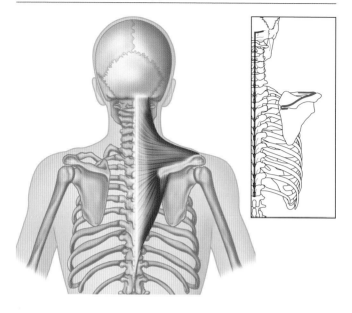

Trapezius

Greek, *trapezoeides*, table shaped.

Origin
Medial third of superior nuchal line of occipital bone. External occipital protuberance. Ligamentum nuchae. Spinous processes and supraspinous ligaments of C7 and T1–12.

Insertion
Superior edge of crest of spine of scapula. Medial border of acromion. Posterior border of lateral one-third of clavicle.

Nerve supply
Motor supply
Accessory nerve (XI).
Sensory supply (proprioception)
Ventral rami of cervical nerves C3 and 4.

Blood supply
Transverse cervical artery (from subclavian artery).

Action
Powerful elevator of the scapula; rotates the scapula during abduction of humerus above horizontal.
Middle fibers retract scapula.
Lower fibers depress scapula, particularly against resistance.

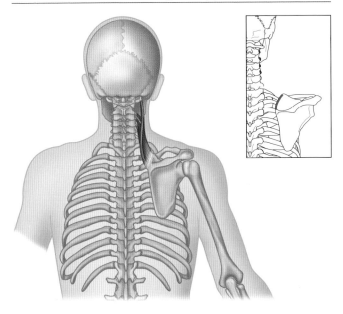

Levator Scapulae

Latin, *levare*, to lift; *scapulae*, of the shoulder blade.

Origin
Transverse processes of C1, 2, and posterior tubercles of transverse processes of C3, 4.

Insertion
Posterior surface of medial border of scapula from superior angle to root of spine of scapula.

Nerve supply
Ventral rami of C3 and C4 spinal nerves and dorsal scapular nerve (C5).

Blood supply
Dorsal scapular artery
via deep branch of transverse cervical artery (from subclavian artery).

Action
Elevates scapula. Helps retract scapula. Helps side flex neck.

Muscles Attaching the Upper Limb to the Trunk

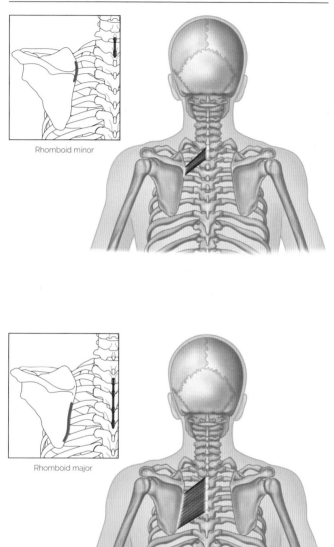

Rhomboid minor

Rhomboid major

Rhomboids

Greek, *rhomboeides*, parallelogram shaped, with only opposite sides and angles equal. **Latin**, *minor*, smaller; *major*, larger.

Origin
Minor
Spinous processes of C7, T1. Lower part of ligamentum nuchae.
Major
Spinous processes of T2–5 and intervening supraspinous ligaments.

Insertion
Minor
Posterior surface of medial border of scapula at the root of spine of scapula.

Major
Posterior surface of medial border of scapula from the root of spine of scapula to the inferior angle.

Nerve supply
Dorsal scapular nerve C4, 5.

Blood supply
Dorsal scapular artery
via deep branch of transverse cervical artery (from subclavian artery).

Action
Elevate and retract scapula.

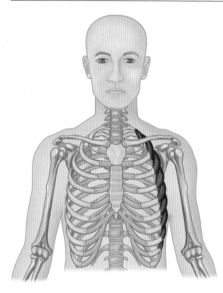

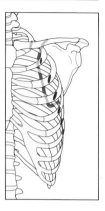

Serratus Anterior

Latin, *serratus*, serrated; *anterior*, at the front.

Origin
Lateral surfaces of upper eight or nine ribs and deep fascia covering the related intercostal spaces.

Insertion
Anterior surface of medial border of scapula.

Nerve supply
Long thoracic nerve C5–7.

Blood supply
Lateral thoracic artery (from axillary artery).

Action
Rotates scapula for abduction and flexion of arm. Protracts scapula.

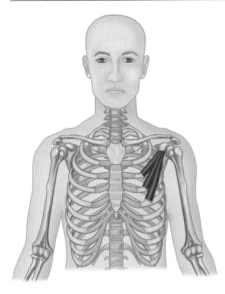
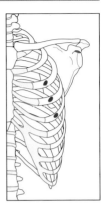

Pectoralis Minor

Latin, *pectoralis*, relating to the chest; *minor*, smaller.

Origin
Outer surfaces of third to fifth ribs, and fascia of the corresponding intercostal spaces.

Insertion
Coracoid process of scapula.

Nerve supply
Medial pectoral nerve C5, (6), 7, 8, T1.

Blood supply
Pectoral branch of thoracoacromial trunk (from axillary artery). Can also be supplied by lateral thoracic artery.

Action
Draws tip of shoulder downward. Protracts scapula. Raises ribs during forced inspiration.

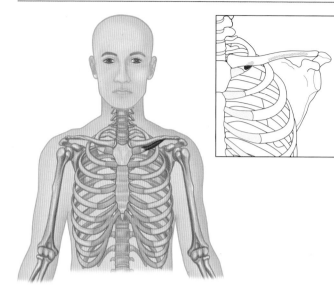

Subclavius

Latin, *sub*, under; *clavis*, key.

Origin
First rib at junction between rib and costal cartilage.

Insertion
Groove on inferior surface of middle one-third of clavicle.

Nerve supply
Nerve to subclavius C5, 6.

Blood supply
Clavicular branch of thoracoacromial trunk (from axillary artery).

Action
Draws tip of shoulder downward. Pulls clavicle medially to stabilize sternoclavicular joint.

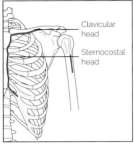

Clavicular head

Sternocostal head

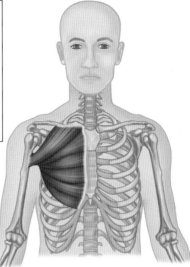

Pectoralis Major

Latin, *pectoralis*, relating to the chest; *major*, larger.

Origin
Clavicular head
Anterior surface of medial half of clavicle.
Sternocostal head
Anterior surface of sternum. First seven costal cartilages. Sternal end of sixth rib. Aponeurosis of external oblique.

Insertion
Lateral lip of intertubercular sulcus of humerus.

Nerve supply
Medial and lateral pectoral nerves: *clavicular head*: C5, 6; *sternocostal head*: C6–8, T1.

Blood supply
Pectoral branch of thoracoacromial trunk and lateral thoracic artery (from axillary artery).

Action
Flexion, adduction, and medial rotation of arm at glenohumeral joint.
Clavicular head
Flexion of extended arm.
Sternocostal head
Extension of flexed arm.

Muscles Attaching the Upper Limb to the Trunk

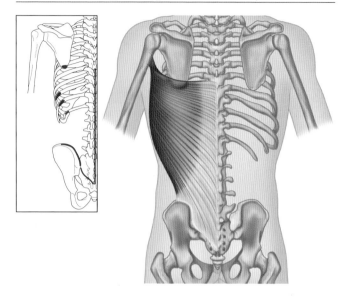

Latissimus Dorsi

Latin, *latissimus*, widest; *dorsi*, of the back.

Origin
Spinous processes of lower six thoracic vertebrae and related interspinous ligaments; via thoracolumbar fascia to the spinous processes of lumbar vertebrae, related interspinous ligaments, and iliac crest. Lower three or four ribs.

Insertion
Twists to insert into the floor of intertubercular sulcus of humerus, just below the shoulder joint.

Nerve supply
Thoracodorsal nerve C6–8.

Blood supply
Thoracodorsal artery
via subscapular artery (from axillary artery).
Dorsal scapular artery
via deep branch of transverse cervical artery (from subclavian artery).

Action
Adduction, medial rotation, and extension of the arm at the glenohumeral joint. It is one of the chief climbing muscles, since it pulls shoulders downward and backward, and pulls trunk up to the fixed arms. Assists in forced inspiration, by raising lower ribs.

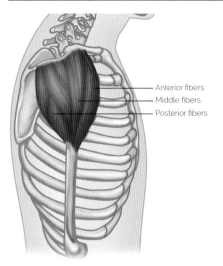

Anterior fibers
Middle fibers
Posterior fibers

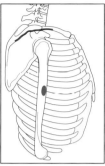

Deltoid

Greek, *deltoeides*, shaped like the Greek capital letter delta (Δ).

Origin
Anterior fibers
Anterior border of lateral one-third of clavicle.
Middle fibers
Lateral margin of acromion process.
Posterior fibers
Inferior edge of crest of spine of scapula.

Insertion
Deltoid tuberosity of humerus.

Nerve supply
Axillary nerve C5, 6.

Blood supply
Posterior circumflex humeral artery and deltoid branch of thoracoacromial artery (from axillary artery).

Action
Major abductor of the arm (abducts arm beyond initial 15 degrees, which is done by supraspinatus); anterior fibers assist in flexing the arm; posterior fibers assist in extending the arm.

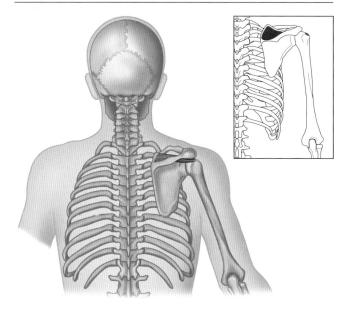

Supraspinatus

Latin, *supra*, above; *spina*, spine.

Origin
Medial two-thirds of supraspinous fossa of scapula and deep fascia that covers the muscle.

Insertion
Most superior facet on the greater tubercle of humerus.

Nerve supply
Suprascapular nerve C5, 6.

Blood supply
Suprascapular artery via thyrocervical trunk
(from subclavian artery).

Action
Initiates abduction of arm to 15 degrees at glenohumeral joint (at which point deltoid takes over).

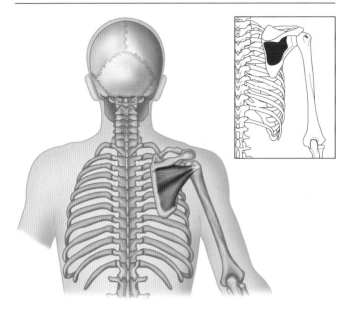

Infraspinatus

Latin, *infra*, below; *spina*, spine.

Origin
Medial two-thirds of infraspinous fossa of scapula and deep fascia that covers the muscle.

Insertion
Middle facet on posterior surface of greater tubercle of humerus.

Nerve supply
Suprascapular nerve C5, 6.

Blood supply
Suprascapular artery
via thyrocervical trunk (from subclavian artery).
Circumflex scapular artery
via subscapular artery (from subclavian artery).

Action
Lateral rotation of arm at glenohumeral joint.

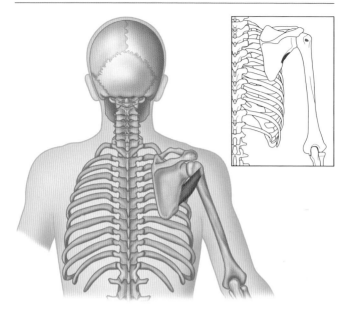

Teres Minor

Latin, *teres*, rounded, finely shaped; *minor*, smaller.

Origin
Upper two-thirds of a strip of bone on posterior surface of scapula immediately adjacent to lateral border of scapula.

Insertion
Inferior facet on greater tubercle of humerus.

Nerve supply
Axillary nerve C5, 6.

Blood supply
Circumflex scapular artery via subscapular artery (from axillary artery).

Action
Lateral rotation of arm at glenohumeral joint.

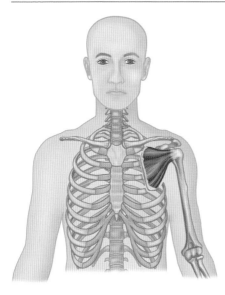

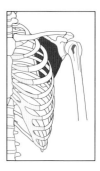

Subscapularis

Latin, *sub*, under; *scapularis*,
relating to the shoulder blade.

Origin
Medial two-thirds of subscapular
fossa.

Insertion
Lesser tubercle of humerus.

Nerve supply
Upper and lower subscapular
nerves C5, 6, (7).

Blood supply
Suprascapular artery
(from axillary artery).

Action
Medial rotation of arm at
glenohumeral joint.

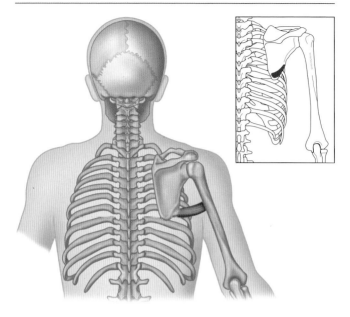

Teres Major

Latin, *teres*, rounded, finely shaped; *major*, larger.

Origin
Oval area on lower third of posterior surface of inferior angle of scapula.

Insertion
Medial lip of intertubercular sulcus on anterior surface of humerus.

Nerve supply
Lower subscapular nerve C5–7.

Blood supply
Circumflex scapular artery via subscapular artery (from axillary artery).

Action
Medial rotation and extension of arm at glenohumeral joint.

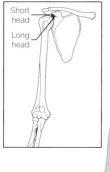

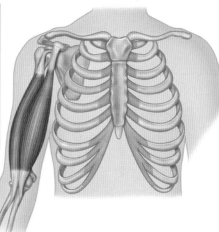

Biceps Brachii

Latin, *biceps*, two-headed; *brachii*, of the arm.

Origin
Long head
Supraglenoid tubercle of scapula.
Short head
Tip of coracoid process.

Insertion
Radial tuberosity.

Nerve supply
Musculocutaneous nerve C5, 6.

Blood supply
Brachial artery
(continuation of axillary artery).

Action
Powerful flexor of forearm at elbow joint. Supinates forearm. Accessory flexor of arm at glenohumeral joint.

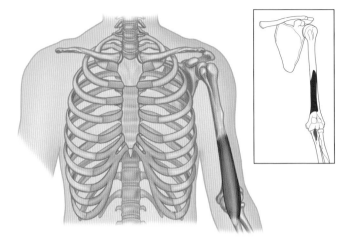

Brachialis

Latin, *brachialis*, relating to the arm.

Origin
Anterior aspect of humerus (medial and lateral surfaces) and adjacent intermuscular septae.

Insertion
Tuberosity of ulna.

Nerve supply
Musculocutaneous nerve C5, 6. Small contribution by radial nerve (C7) to lateral part of muscle.

Blood supply
Brachial artery (continuation of axillary artery). **Radial recurrent artery** (from radial artery).

Action
Powerful flexor of forearm at elbow joint.

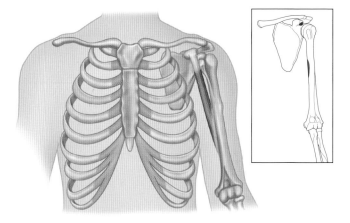

Coracobrachialis

Greek, *korakoeides*, raven-like.
Latin, *brachialis*, relating to
the arm.

Blood supply
Brachial artery
(continuation of axillary artery).

Origin
Tip of coracoid process.

Action
Flexor of arm at glenohumeral
joint.

Insertion
Medial aspect of humerus at
mid-shaft.

Nerve supply
Musculocutaneous nerve C5–7.

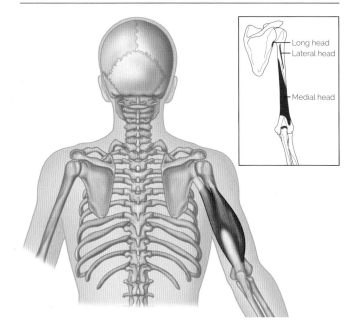

Triceps Brachii

Latin, *triceps*, three-headed; *brachii*, of the arm.

Origin

Long head
Infraglenoid tubercle of scapula.
Medial head
Posterior surface of humerus (below and medial to radial groove).
Lateral head
Posterior surface of humerus (above and lateral to radial groove).

Insertion
Posterior part of olecranon process of ulna.

Nerve supply
Radial nerve C6–8.

Blood supply
Deep brachial artery
(via brachial artery continuing from axillary artery).

Action
Extends forearm at elbow joint. Long head can also extend and adduct arm at shoulder joint.

CHAPTER 7

Muscles of the Forearm and Hand

The forearm, which lies between the elbow joint and the wrist joint, contains two long, parallel bones—the ulna and the radius. The ulna is the medial bone of the forearm and is the longer and larger of the two bones; the radius is slightly shorter, thinner, and located on the lateral side of the forearm. At the elbow, the proximal end of the ulna is at its widest and the proximal end of the radius at its narrowest. This difference is reversed at the distal ends: the radius widens to make up the bulk of the wrist joint, together with the now much narrower ulnar and carpal bones.

In cross section, the forearm can be divided into anterior and posterior compartments.

The anterior compartment contains the **forearm flexors**, arranged in superficial, intermediate, and deep strata.

Following a similar pattern of superficial and deep layers, but this time arising from the lateral epicondyle, the posterior compartment contains the **extensors** of the wrist and fingers, which act as antagonists to the flexor muscles. In general, the extensors are somewhat weaker than the flexor muscles that they work against.

The principal role of the hand itself is grasping and manipulation, with the muscle groups involved termed **intrinsic** and **extrinsic**. The **extrinsic muscles** originate more proximally in the forearm and insert into the hand as long tendons to provide crude movements. The **intrinsic muscles**, located within the hand itself, are responsible for fine control of the complicated movements of the fingers.

The muscles of the thenar and hypothenar eminences have a role in coordinating numerous movements of the thumb and little finger, including their opposition to one another.

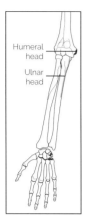

Humeral head
Ulnar head

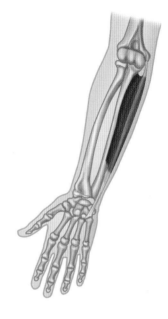

Flexor Carpi Ulnaris

Latin, *flectere*, to bend; *carpi*, of the wrist; *ulnaris*, relating to the elbow / arm.

Origin
Humeral head
Medial epicondyle of humerus.
Ulnar head
Medial border of olecranon and posterior border of upper two-thirds of ulna.

Insertion
Pisiform bone. Hook of hamate. Base of fifth metacarpal.

Nerve supply
Ulnar nerve C7, 8, T1.

Blood supply
Ulnar artery
(from brachial artery).

Action
Flexes and adducts wrist.

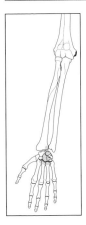

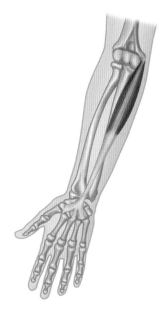

Palmaris Longus

Latin, *palmaris*, relating to the palm; *longus*, long.

Origin
Medial epicondyle of humerus.

Insertion
Palmar aponeurosis of hand.

Nerve supply
Median nerve C(6), 7, 8.

Blood supply
Ulnar artery
(from brachial artery).

Action
Flexes wrist joint. Tenses palmar fascia.

Muscles of the Anterior Compartment of the Forearm—Superficial Layer

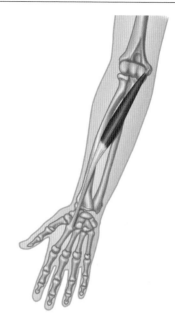

Flexor Carpi Radialis

Latin, *flectere*, to bend; *carpi*, of the wrist; *radius*, staff, spoke of wheel.

Origin
Medial epicondyle of humerus.

Insertion
Bases of second and third metacarpals.

Median nerve C6, 7.

Blood supply
Ulnar and radial arteries (from brachial artery).

Action
Flexes and abducts wrist joint.

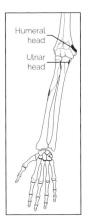

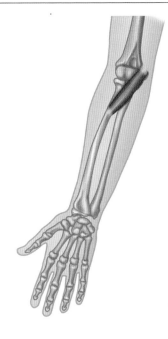

Pronator Teres

Latin, *pronare*, to bend forward; *teres*, rounded, finely shaped.

Origin
Humeral head
Medial epicondyle and adjacent supra-epicondylar ridge.
Ulnar head
Medial border of coronoid process.

Insertion
Mid-lateral surface of radius (pronator tuberosity).

Nerve supply
Median nerve C6, 7.

Blood supply
Ulnar artery
(from brachial artery).
Anterior ulnar recurrent branch of the ulnar artery

Action
Pronates forearm.

Muscles of the Anterior Compartment of the Forearm—Intermediate Layer

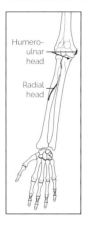

Humero-ulnar head

Radial head

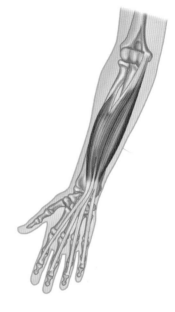

Flexor Digitorum Superficialis

Latin, *flectere*, to bend; *digitorum*, of the fingers; *superficialis*, on the surface.

Origin
Humero-ulnar head
Medial epicondyle of humerus. Adjacent border of coronoid process.
Radial head
Oblique line of radius.

Insertion
Four tendons each divide into two slips, each of which insert into the sides of the middle phalanges of the four fingers.

Median nerve C8, T1.

Blood supply
Ulnar artery
(from brachial artery).

Action
Flexes proximal interphalangeal joints of the index, middle, ring, and little fingers; can also flex metacarpophalangeal joints of the same fingers and the wrist joint.

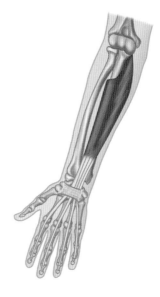

Flexor Digitorum Profundus

Latin, *flectere*, to bend; *digitorum*, of the fingers; *profundus*, deep.

Origin
Medial and anterior surfaces of ulna. Medial half of interosseous membrane.

Insertion
Four tendons, which attach to the palmar surfaces of the distal phalanges of the index, middle, ring, and little fingers.

Nerve supply
Medial half of muscle, destined for the little and ring fingers
Ulnar nerve C8, T1.

Lateral half of muscle, destined for the index and middle fingers
Anterior interosseous branch of median nerve C8, T1.

Blood supply
Ulnar artery
(from brachial artery).
Anterior interosseous artery
(from ulnar artery).

Action
Flexes distal interphalangeal joints of the index, middle, ring, and little fingers; can also flex metacarpophalangeal joints of the same fingers and the wrist joint.

Muscles of the Anterior Compartment of the Forearm—Deep Layer

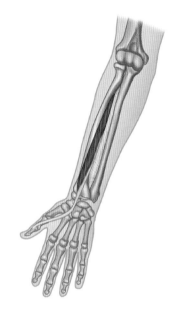

Flexor Pollicis Longus

Latin, *flectere*, to bend; *pollicis*, of the thumb; *longus*, long.

Origin
Anterior surface of shaft of radius. Radial half of interosseous membrane.

Insertion
Palmar surface of base of distal phalanx of thumb.

Nerve supply
Anterior interosseous branch of median nerve C(6), 7, 8.

Blood supply
Anterior interosseous artery (from ulnar artery).

Action
Flexes interphalangeal joint of thumb. Assists in flexion of metacarpophalangeal joint of thumb.

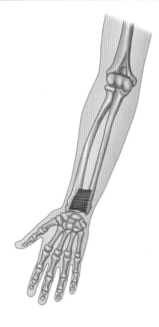

Pronator Quadratus

Latin, *pronare*, to bend forward; *quadratus*, squared.

Origin
Linear ridge on distal anterior surface of ulna.

Insertion
Distal anterior surface of radius.

Nerve supply
Anterior interosseous branch of median nerve C7, 8.

Blood supply
Anterior interosseous artery (from ulnar artery).

Action
Pronates forearm and hand. Helps hold radius and ulna together, reducing stress on inferior radioulnar joint.

Muscles of the Posterior Compartment of the Forearm—Superficial Layer

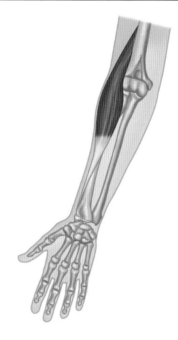

Brachioradialis

Latin, *brachium*, arm; *radius*, staff, spoke of wheel.

Origin
Proximal part of lateral supraepicondylar ridge of humerus and adjacent intermuscular septum.

Insertion
Lower surface of distal end of radius, just above styloid process.

Nerve supply
Radial nerve C5, 6.

Blood supply
Radial recurrent branch of radial artery (from brachial artery).

Action
Accessory flexor of elbow joint when forearm is midpronated.

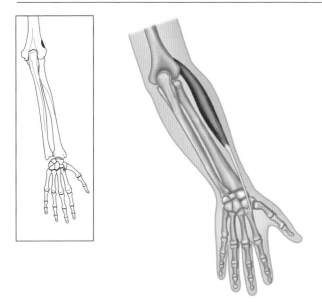

Extensor Carpi Radialis Longus

Latin, *extendere*, to extend; *carpi*, of the wrist; *radius*, staff, spoke of wheel; *longus*, long.

Origin
Distal part of lateral supraepicondylar ridge of humerus and adjacent intermuscular septum.

Insertion
Dorsal surface of base of second metacarpal.

Nerve supply
Radial nerve C6, 7.

Blood supply
Radial artery
(from brachial artery).

Action
Extends and abducts wrist.

Muscles of the Posterior Compartment of the Forearm—Superficial Layer

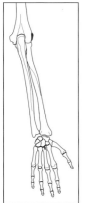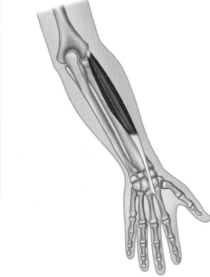

Extensor Carpi Radialis Brevis

Latin, *extendere*, to extend; *carpi*, of the wrist; *radius*, staff, spoke of wheel; *brevis*, short.

Origin
Lateral epicondyle of humerus and adjacent intermuscular septum.

Insertion
Dorsal surface of base of second and third metacarpals.

Nerve supply
Radial nerve C7, 8.

Blood supply
Radial artery
(from brachial artery).

Action
Extends and abducts wrist.

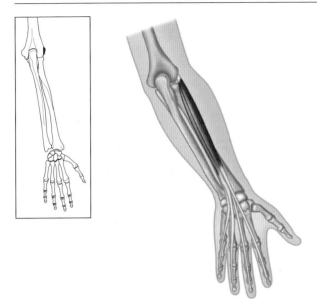

Extensor Digitorum

Latin, *extendere*, to extend; *digitorum*, of the fingers.

Origin
Lateral epicondyle of humerus and adjacent intermuscular septum and deep fascia.

Insertion
Four tendons, which insert via extensor hoods into the dorsal aspects of the bases of the middle and distal phalanges of the index, middle, ring, and little fingers.

Nerve supply
Posterior interosseous nerve C7, 8.

Blood supply
Recurrent interosseous and posterior interosseous arteries via common interosseous artery (from ulnar artery).

Action
Extends the index, middle, ring, and little fingers; can also extend the wrist.

Muscles of the Posterior Compartment of the Forearm—Superficial Layer

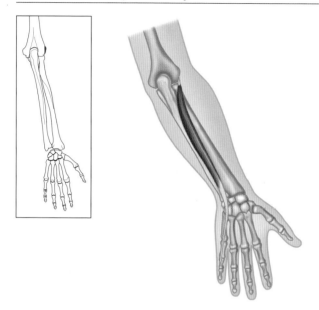

Extensor Digiti Minimi

Latin, *extendere*, to extend; *digiti*, of the finger; *minimi*, of the smallest.

Origin
Lateral epicondyle of humerus and adjacent intermuscular septum together with extensor digitorum.

Insertion
Extensor hood of little finger.

Nerve supply
Posterior interosseous nerve C6, 7, 8.

Blood supply
Recurrent interosseous artery via common interosseous artery (from ulnar artery).

Action
Extends little finger.

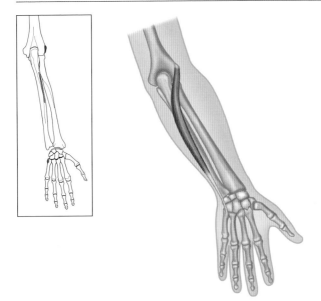

Extensor Carpi Ulnaris

Latin, *extendere*, to extend; *carpi*, of the wrist; *ulnaris*, relating to the elbow/arm.

Origin
Lateral epicondyle of humerus and posterior border of ulna.

Insertion
Tubercle on base of medial side of fifth metacarpal.

Nerve supply
Posterior interosseous nerve C6, 7, 8.

Blood supply
Ulnar artery
(from brachial artery).

Action
Extends and adducts wrist.

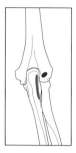

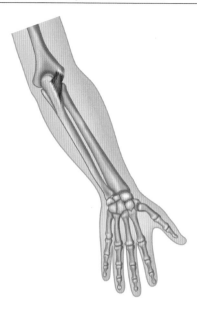

Anconeus

Greek, *agkon*, elbow.

Origin
Lateral epicondyle of humerus.

Insertion
Lateral surface of olecranon process and proximal posterior surface of ulna.

Nerve supply
Radial nerve C6, 7, 8.

Blood supply
Middle collateral branch of deep brachial artery (from ulnar artery).
Recurrent interosseous artery via common interosseous artery (from ulnar artery).

Action
Abduction of ulna in pronation. Accessory extensor of elbow joint.

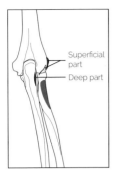

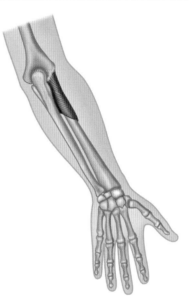

Supinator

Latin, *supinus*, lying on the back.

Origin
Superficial part
Lateral epicondyle of humerus.
Radial collateral and anular
ligaments.
Deep part
Supinator crest of ulna.

Insertion
Lateral surface of radius superior
to the anterior oblique line.

Nerve supply
Posterior interosseous nerve C5,
6, (7).

Blood supply
Recurrent interosseous artery
via common interosseous artery
(from ulnar artery).
Occasionally also supplied by
recurrent radial artery.

Action
Supination.

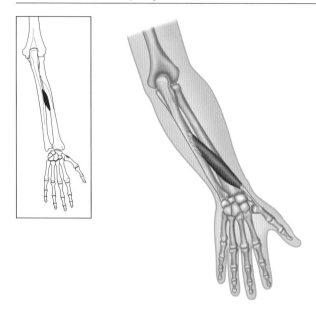

Abductor Pollicis Longus

Latin, *abducere*, to lead away from; *pollicis*, of the thumb; *longus*, long.

Origin
Posterior surfaces of ulna and radius, distal to attachments of supinator and anconeus. Intervening interosseous membrane.

Insertion
Lateral side of base of first metacarpal.

Nerve supply
Posterior interosseous nerve C7, 8.

Blood supply
Posterior interosseous artery via common interosseous artery (from ulnar artery).

Action
Abducts carpometacarpal joint of thumb; accessory extensor of thumb.

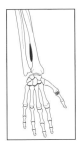

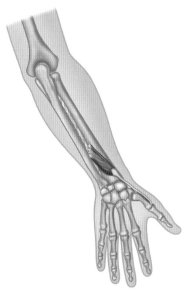

Extensor Pollicis Brevis

Latin, *extendere*, to extend; *pollicis*, of the thumb; *brevis*, short.

Origin
Posterior surface of radius, distal to origin of abductor pollicis longus. Adjacent interosseous membrane.

Insertion
Base of dorsal surface of proximal phalanx of thumb.

Nerve supply
Posterior interosseous nerve C7, 8.

Blood supply
Posterior interosseous artery via common interosseous artery (from ulnar artery).

Action
Extends metacarpophalangeal joint of thumb. Can also extend carpometacarpal joint of thumb.

Muscles of the Posterior Compartment of the Forearm—Deep Layer

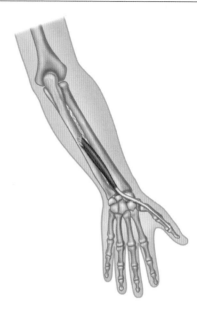

Extensor Pollicis Longus

Latin, *extendere*, to extend; *pollicis*, of the thumb; *longus*, long.

Origin
Posterior surface of ulna, distal to abductor pollicis longus. Adjacent interosseous membrane.

Insertion
Dorsal surface of base of distal phalanx of thumb.

Nerve supply
Posterior interosseous nerve C7, 8.

Blood supply
Posterior interosseous artery via common interosseous artery (from ulnar artery).

Action
Extends interphalangeal joint of thumb. Can also extend carpometacarpal and metacarpophalangeal joint of thumb.

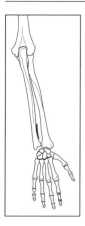

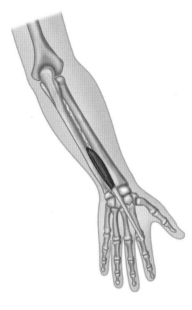

Extensor Indicis

Latin, *extendere*, to extend; *indicis*, of the index finger.

Origin
Posterior surface of ulna, distal to extensor pollicis longus. Adjacent interosseous membrane.

Insertion
Extensor hood of index finger.

Nerve supply
Posterior interosseous nerve C7, 8.

Blood supply
Posterior interosseous artery via common interosseous artery (from ulnar artery).

Action
Extends index finger.

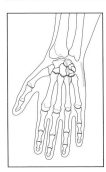

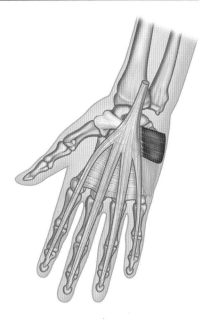

Palmaris Brevis

Latin, *palmaris*, relating to the palm; *brevis*, short.

Origin
Palmar aponeurosis. Flexor retinaculum.

Insertion
Skin on ulnar border of hand.

Nerve supply
Superficial branch of ulnar nerve C(7), 8, T1.

Blood supply
Ulnar artery
(from brachial artery).

Action
Improves grip.

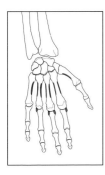

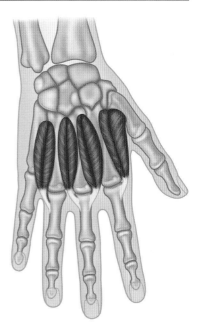

Dorsal Interossei

Latin, *dorsalis*, relating to the back; *interosseus*, between bones.

Origin
By two heads, each from adjacent sides of metacarpals.

Insertion
Extensor hood and base of proximal phalanges of index, middle, and ring fingers.

Nerve supply
Deep branch of ulnar nerve C8, T1.

Blood supply
Dorsal metacarpal arteries and palmar metacarpal arteries of deep palmar arch

Action
Abduction of index, middle, and ring fingers at metacarpophalangeal joints.

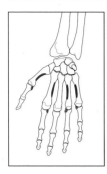

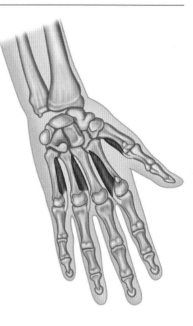

Palmar Interossei

Latin, *palmaris*, relating to the palm; *interosseus*, between bones.

Origin
Sides of metacarpals.

Insertion
Extensor hoods of the thumb, index, ring, and little fingers and proximal phalanx of thumb.

Nerve supply
Deep branch of ulnar nerve C8, T1.

Blood supply
Palmar metacarpal arteries of deep palmar arch

Action
Adduction of the thumb, index, ring, and little fingers at metacarpophalangeal joints.

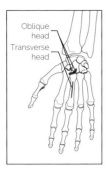

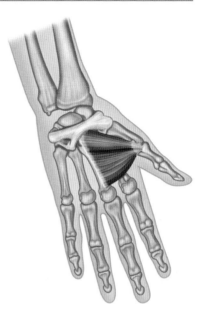

Adductor Pollicis

Latin, *adducere,* to lead to; *pollicis,* of the thumb.

Origin
Transverse head
Palmar surface of third metacarpal.
Oblique head
Capitate and bases of second and third metacarpals.

Insertion
Base of proximal phalanx of thumb and extensor hood of thumb.

Nerve supply
Deep branch of ulnar nerve C8, T1.

Blood supply
Palmar metacarpal arteries of deep palmar arterial arch (from radial artery).

Action
Adducts thumb.

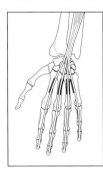

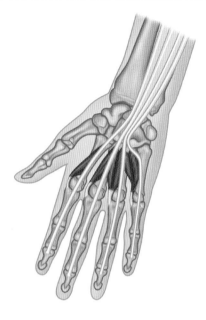

Lumbricals

Latin, *lumbricus*, earthworm.

Origin
Tendons of flexor digitorum profundus.

Insertion
Extensor hoods of index, ring, middle, and little fingers.

Nerve supply
Lateral lumbricals (first and second)
Digital branches of median nerve.

Medial lumbricals (third and fourth)
Deep branch of ulnar nerve.

Blood supply
Palmar metacarpal arteries of deep palmar arch

Action
Extend interphalangeal joints and simultaneously flex metacarpophalangeal joints.

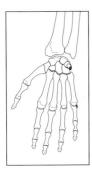

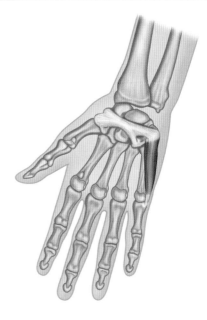

Abductor Digiti Minimi

Latin, *abducere*, to lead away from; *digiti*, of the finger; *minimi*, of the smallest.

Origin
Pisiform, pisohamate ligament, and tendon of flexor carpi ulnaris.

Insertion
Proximal phalanx of little finger.

Nerve supply
Deep branch of ulnar nerve C(7), 8, T1.

Blood supply
Deep palmar branches of ulnar artery
(from brachial artery).

Action
Abducts little finger at metacarpophalangeal joint. A surprisingly powerful muscle, which particularly comes into play when fingers are spread to grasp a large object.

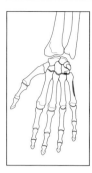

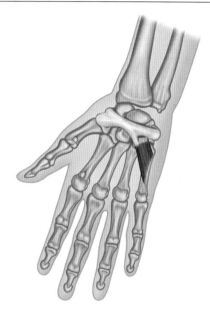

Opponens Digiti Minimi

Latin, *opponens*, opposing; *digiti*, of the finger; *minimi*, of the smallest.

Origin
Hook of hamate. Flexor retinaculum.

Insertion
Entire length of medial (ulnar) border of fifth metacarpal.

Nerve supply
Deep branch of ulnar nerve C(7), 8, T1.

Blood supply
Deep palmar branches of ulnar artery
(from brachial artery).

Action
Laterally rotates fifth metacarpal.

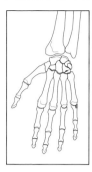

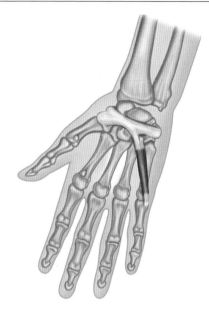

Flexor Digiti Minimi Brevis

Latin, *flectere*, to flex; *digiti*, of the finger; *minimi*, of the smallest; *brevis*, short.

Origin
Hook of hamate. Flexor retinaculum.

Insertion
Proximal phalanx of little finger.

Nerve supply
Deep branch of ulnar nerve C(7), 8, T1.

Blood supply
Ulnar artery
(from brachial artery).

Action
Flexes little finger at metacarpophalangeal joint.

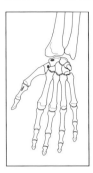

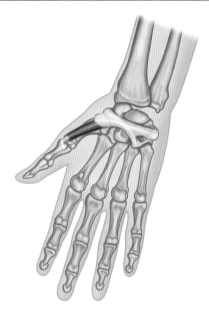

Abductor Pollicis Brevis

Latin, *abducere*, to lead away from; *pollicis*, of the thumb; *brevis*, short.

Origin
Tubercles of trapezium and scaphoid and adjacent flexor retinaculum.

Insertion
Proximal phalanx and extensor hood of thumb.

Nerve supply
Recurrent branch of median nerve C8, T1.

Blood supply
Superficial palmar branches of radial artery
(from brachial artery).

Action
Abducts thumb at metacarpophalangeal joint.

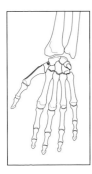

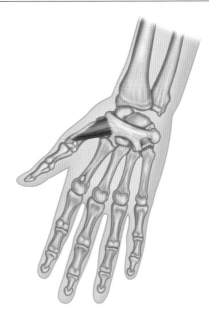

Opponens Pollicis

Latin, *opponens*, opposing; *pollicis*, of the thumb.

Origin
Flexor retinaculum. Tubercle of trapezium.

Insertion
Entire length of radial border of first metacarpal.

Nerve supply
Recurrent branch of median nerve C8, T1.

Blood supply
Superficial palmar branches of radial artery
(from brachial artery).

Action
Medially rotates thumb.

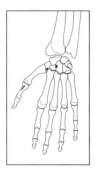

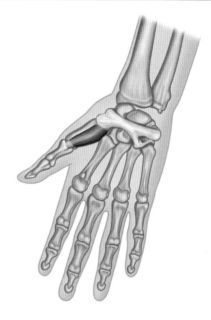

Flexor Pollicis Brevis

Latin, *flectere*, to flex; *pollicis*, of the thumb; *brevis*, short.

Origin
Flexor retinaculum. Tubercle of trapezium.

Insertion
Proximal phalanx of thumb.

Nerve supply
Recurrent branch of median nerve C8, T1.

Blood supply
Superficial palmar branches of radial artery
(from brachial artery).

Action
Flexes thumb at metacarpophalangeal joint.

CHAPTER 8

Muscles of the Hip and Thigh

The muscles of the hip and thigh provide not only stability but movement and strength too; depending upon their locations and functions, these muscles can be divided up into four groups—anterior, adductor, abductor, and posterior.

The **anterior muscle group**, responsible for flexing the thigh at the hip, includes:

- **Iliopsoas**, consisting of two muscles: **psoas major** and **iliacus** (see Chapter 5).
- **Quadriceps femoris**, consisting of four muscles (the name means four-headed): **rectus femoris**, **vastus intermedius**, **vastus lateralis**, and **vastus medialis**.

The **adductor muscle group**, on the medial side of the thigh, includes:

- **Adductor longus**, **adductor brevis**, **adductor magnus**, **pectineus**, and **gracilis**.

The **abductor muscle group**, on the lateral side of the thigh, includes:

- **Piriformis**, **gemellus superior**, **gemellus inferior**, **tensor fasciae latae**, **sartorius**, **gluteus medius**, and **gluteus minimus**.

The **posterior muscle group**, includes:

- **Gluteus maximus** (the largest muscle in the body).
- **Hamstrings**, consisting of three muscles: **biceps femoris**, **semimembranosus**, and **semitendinosus**.

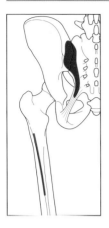

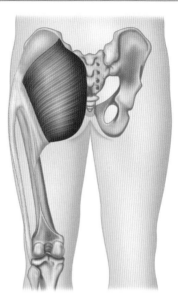

Gluteus Maximus

Greek, *gloutos*, buttock. *Latin*, *maximus*, biggest.

Origin
Fascia covering gluteus medius, external surface of ilium behind posterior gluteal line, fascia of erector spinae, dorsal surface of lower sacrum, lateral margin of coccyx, external surface of sacrotuberous ligament.

Insertion
Posterior aspect of iliotibial tract of fascia lata. Gluteal tuberosity of proximal femur.

Nerve supply
Inferior gluteal nerve L5, S1, 2.

Blood supply
Inferior and superior gluteal arteries
via internal iliac artery (a branch of common iliac artery from abdominal aorta).
First perforating branch of the deep femoral artery
(via external iliac artery).

Action
Powerful extensor of flexed femur at hip joint. Lateral stabilizer of hip and knee joints. Laterally rotates and abducts thigh.

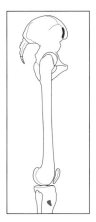

Tensor Fasciae Latae

Latin, *tendere*, to stretch, pull; *fascia*, band; *lata*, side or lateral.

Origin
Lateral aspect of crest of ilium between ASIS and tubercle of the crest.

Insertion
Iliotibial tract which inserts into the upper lateral tibia.

Nerve supply
Superior gluteal nerve L4, 5, S1.

Blood supply
Superior gluteal artery
via internal iliac artery (a branch of common iliac artery from abdominal aorta).
Lateral circumflex femoral artery
via deep femoral artery (from external iliac artery).

Action
Stabilizes the knee in extension.

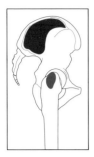

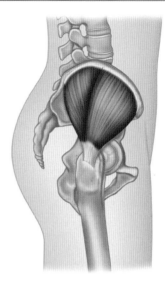

Gluteus Medius

Greek, *gloutos,* buttock. **Latin,** *medius,* middle.

Origin
External surface of ilium between anterior and posterior gluteal lines.

Insertion
Oblique ridge on lateral surface of greater trochanter.

Nerve supply
Superior gluteal nerve L4, 5, S1.

Blood supply
Superior gluteal artery
via internal iliac artery (a branch of common iliac artery from abdominal aorta).

Action
Abducts femur at hip joint. Medially rotates thigh. Holds pelvis secure over stance leg and prevents pelvic drop on the opposite swing side during walking (Trendelenburg gait).

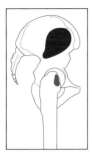

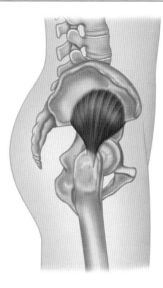

Gluteus Minimus

Greek, *gloutos*, buttock. **Latin**, *minimus*, smallest.

Origin
External surface of ilium between anterior and inferior gluteal lines.

Insertion
Anterolateral border of greater trochanter.

Nerve supply
Superior gluteal nerve L4, 5, S1.

Blood supply
Superior gluteal artery
via internal iliac artery (a branch of common iliac artery from abdominal aorta).

Action
Abducts, medially rotates, and may assist in flexion of hip joint.

Muscles of the Gluteal Region—Deep Lateral Hip Rotators

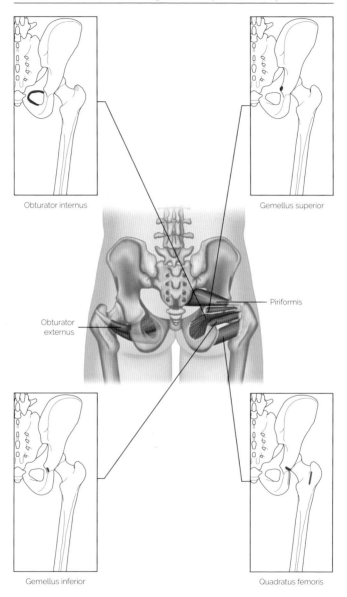

Obturator internus

Gemellus superior

Piriformis

Obturator externus

Gemellus inferior

Quadratus femoris

Latin, *obturare*, to obstruct; *internus*, internal, *gemellus*, twin/double; *superior*, upper; *inferior*, lower; *quadratus*, squared; *femoris*, of the thigh.

Origin

Obturator internus
Anterolateral wall of true pelvis; deep surface of obturator membrane and surrounding bone.
Gemellus superior
External surface of ischial spine.
Gemellus inferior
Upper aspect of ischial tuberosity.
Quadratus femoris
Lateral edge of ischium just anterior to ischial tuberosity.

Insertion

Obturator internus
Medial side of greater trochanter.
Gemellus superior
Along length of superior surface of obturator internus tendon and into medial side of greater trochanter with obturator internus tendon.
Gemellus inferior
Along length of inferior surface of obturator internus tendon and into medial side of greater trochanter with obturator internus tendon.
Quadratus femoris
Quadrate tubercle on intertrochanteric crest of proximal femur.

Nerve supply

Obturator internus and gemellus superior
Nerve to obturator internus, L5, S1.
Gemellus inferior and quadratus femoris
Nerve to quadratus femoris, L5, S1, (2).

Blood supply

Obturator internus
Inferior and superior gluteal arteries and obturator artery
via internal iliac artery (a branch of common iliac artery from abdominal aorta).
Gemelli
Inferior gluteal artery
via internal iliac artery (a branch of common iliac artery from abdominal aorta).
Quadratus femoris
Obturator artery
via internal iliac artery (a branch of common iliac artery from abdominal aorta), plus can also be supplied by medial circumflex arteries (from deep femoral artery).

Action

Laterally rotate hip joint.
Abduct flexed femur at hip joint.
Help hold head of femur in acetabulum.

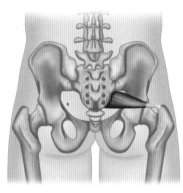

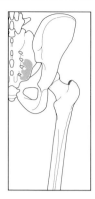

Piriformis

Latin, *pirum*, pear; *forma*, shape.

Piriformis leaves the pelvis by passing through the greater sciatic foramen, and along with obturator internus, is a muscle of the pelvic wall.

Origin
Anterior surface of sacrum between anterior sacral foramina.

Insertion
Medial side of superior border of greater trochanter.

Nerve supply
Branches from sacral nerves S1, 2.

Blood supply
Inferior and superior gluteal arteries
via internal iliac artery (a branch of common iliac artery from abdominal aorta).

Action
Laterally rotates extended femur at hip joint. Abducts flexed femur at hip joint. Helps hold head of femur in acetabulum. May assist with medial rotation when hip is flexed to 90 degrees and beyond.

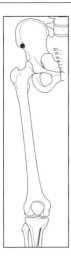

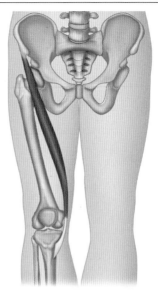

Sartorius

Latin, *sartor*, tailor.

Sartorius is the most superficial muscle of the anterior compartment of the thigh and is also the longest strap muscle in the body. The medial border of the upper third of this muscle forms the lateral boundary of the femoral triangle (adductor longus forms the medial boundary, and the inguinal ligament forms the superior boundary). The action of sartorius is to put the lower limbs in the seated cross-legged position of the tailor (hence its name from the Latin).

Origin
Anterior superior iliac spine.

Insertion
Medial surface of tibia just inferomedial to tibial tuberosity.

Nerve supply
Femoral nerve L2, 3, (4).

Blood supply
Lateral circumflex femoral artery (from deep femoral artery).
Saphenous branch of descending genicular artery (from femoral artery).

Action
Flexes the thigh at the hip joint (helping to bring leg forward in walking or running). Flexes the leg at the knee joint.

Muscles of the Anterior Compartment of the Thigh—Quadriceps

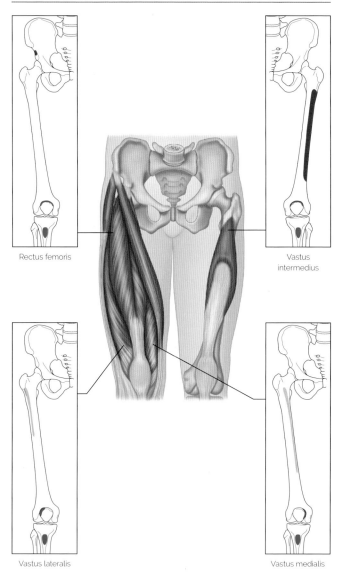

Rectus femoris

Vastus intermedius

Vastus lateralis

Vastus medialis

Latin, *rectus*, straight; *femoris*, of the thigh; *vastus*, vast; *lateralis*, relating to the side.

The four quadriceps (**Latin**: four-headed) femoris muscles are: rectus femoris, vastus lateralis, vastus medialis, and vastus intermedius.

Vastus intermedius is the deepest part of the quadriceps femoris. This muscle has a membranous tendon on its anterior surface to allow a gliding movement between itself and the rectus femoris that overlies it.

Origin
Rectus femoris
Straight head (anterior head): anterior inferior iliac spine; reflected head (posterior head): groove above acetabulum (on ilium).
Vasti group
Upper half of shaft of femur.

Insertion
Patella, then via patellar ligament, into the tibial tuberosity.

Nerve supply
Femoral nerve L2, 3, 4.

Blood supply
Lateral circumflex femoral artery (from deep femoral artery). Plus:
Lateralis only
Can also be supplied by perforating branches of the deep femoral artery
Medialis only
Saphenous branch of descending genicular artery (from femoral artery).
Insertion also supplied by the medial superior genicular branch of popliteal artery (a continuation of femoral artery).

Action
Rectus femoris
Flexes the thigh at the hip joint (particularly in combination, as in kicking a ball), and extends leg at the knee joint.
Vasti group
Extend leg at the knee joint.

Muscles of the Medial Compartment of the Thigh

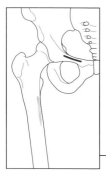

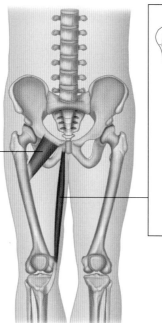

Pectineus

Gracilis

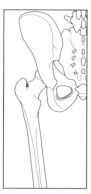

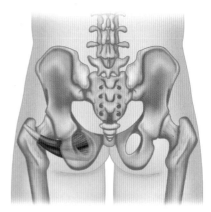

Obturator externus

Muscles of the Medial Compartment of the Thigh

Gracilis

Latin, *gracilis*, slender, delicate.

Origin
A line on the external surfaces of the pubis, the inferior pubic ramus, and ramus of the ischium.

Insertion
Medial surface of proximal shaft of tibia.

Nerve supply
Obturator nerve L2, 3.

Blood supply
Obturator artery
via internal iliac artery (a branch of common iliac artery from abdominal aorta).
Can also be supplied by medial circumflex femoral artery (from deep femoral artery).

Action
Adducts thigh at hip joint. Flexes leg at knee joint.

Pectineus

Latin, *pecten*, comb; *pectinatus*, comb shaped.

Origin
Pecten pubis and adjacent bone of pelvis.

Insertion
Oblique line, from base of lesser trochanter to linea aspera of femur.

Nerve supply
Femoral nerve L2, 3.

Blood supply
Medial circumflex femoral artery
(from deep femoral artery).

Action
Adducts and flexes thigh at hip joint.

Obturator Externus

Latin, *obturare*, to obstruct; *externus*, external.

Origin
External surface of obturator membrane and adjacent bone.

Insertion
Trochanteric fossa.

Nerve supply
Posterior division of obturator nerve L3, 4.

Blood supply
Obturator artery
via internal iliac artery (a branch of common iliac artery from abdominal aorta), plus can also be supplied by medial circumflex arteries (from deep femoral artery).

Action
Laterally rotates thigh at hip joint.

Muscles of the Medial Compartment of the Thigh—Adductors

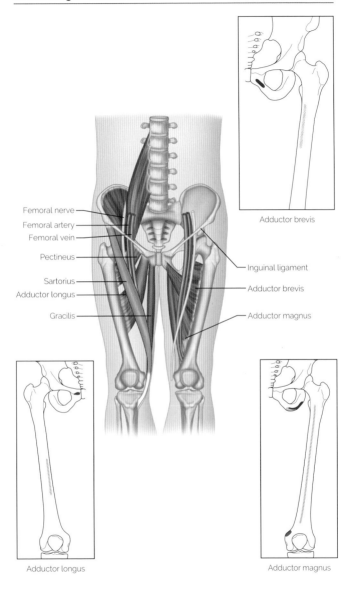

Adductor brevis

Femoral nerve
Femoral artery
Femoral vein
Pectineus
Sartorius
Adductor longus
Gracilis

Inguinal ligament
Adductor brevis
Adductor magnus

Adductor longus

Adductor magnus

Latin, *adducere*, to lead to; *magnus*, large; *brevis*, small; *longus*, long.

Adductor magnus is the largest of the adductor muscle group, which also includes adductor brevis and adductor longus. Its upper fibers are often fused with those of quadratus femoris. Adductor longus is the most anterior of the three.

The lateral border of the upper fibers of adductor longus form the medial border of the **femoral triangle** (sartorius forms the lateral boundary; the inguinal ligament forms the superior boundary).

Origin
Anterior part of the pubic bone (ramus). Adductor magnus also takes its origin from the ischial tuberosity.

Insertion
Entire length of femur, along linea aspera and medial supracondylar line to adductor tubercle on medial epicondyle of femur.

Nerve supply
Magnus
Obturator nerve L2, 3, 4. Sciatic nerve (tibial division) L2, 3, 4.
Brevis
Obturator nerve L2, 3.
Longus
Obturator nerve (anterior division) L2, 3, 4.

Blood supply
Obturator artery
via internal iliac artery (a branch of common iliac artery from abdominal aorta).
Deep femoral artery *(brevis and longus only)*
Medial circumflex femoral artery
(from deep femoral artery).

Action
Adduct and medially rotate thigh at hip joint.

Muscles of the Posterior Compartment of the Thigh—Hamstrings

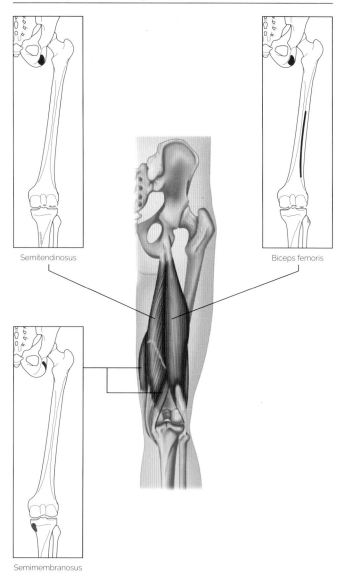

Semitendinosus

Biceps femoris

Semimembranosus

Latin, *semi*, half; *membranosus*, membranous; *tendinosus*, tendinous; *biceps*, two-headed; *femoris*, of the thigh.

The hamstrings consist of three muscles; from medial to lateral: semimembranosus, semitendinosus, and biceps femoris.

Origin
Ischial tuberosity. Biceps femoris (short head only): lateral lip of linea aspera.

Insertion
Semimembranosus
Groove and adjacent bone on medial and posterior surface of medial tibial condyle.
Semitendinosus
Medial surface of proximal tibia.
Biceps femoris
Head of fibula.

Nerve supply
Sciatic nerve L5, S1, 2.

Blood supply
Perforating branches of the deep femoral artery
via femoral artery (a continuation of the external iliac artery).
Inferior gluteal artery via internal iliac artery
(a branch of common iliac artery from abdominal aorta).

Action
Flex leg at knee joint. Semimembranosus and semitendinosus extend thigh at hip joint, medially rotate thigh at hip joint and leg at knee joint. Biceps femoris extends and laterally rotates thigh at hip joint and laterally rotates leg at knee joint.

CHAPTER 9

Muscles of the Leg and Foot

Muscles of the Leg

With the exception of popliteus, all muscles in the leg are attached to the foot; they may be classified, depending on their position, into anterior, posterior, and lateral groups. In addition, the posterior group may be further subdivided into superficial, intermediate, and deep layers.

The **anterior group** of muscles, at the front of the leg, is also called the *extensor compartment*; it functionally extends (dorsiflexes) the foot at the ankle joint and extends the toes. There are four muscles within this group.

Muscles of the **posterior group**, in the calf region of the leg, are concerned with plantar flexion of the ankle and flexion of the toes. The muscles of this *flexor compartment* are arranged in three layers like an onion skin. In the most superficial layer lie the flexor muscles of the ankle comprising a pair of muscles, **gastrocnemius** and **soleus** also known as triceps surae, and plantaris. The intermediate layer comprises **flexor hallucis longus** and **flexor digitorum longus**. The **tibialis posterior** and **popliteus** muscles form the deepest of these layers.

Situated in the *lateral compartment* are: **fibularis longus**, arising from the upper part of the fibula, and **fibularis brevis** from lower down the fibula.

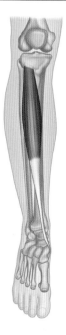

Tibialis Anterior

Latin, *tibialis*, relating to the shin; *anterior*, at the front.

Origin
Lateral surface of tibia and adjacent interosseous membrane.

Insertion
Medial and inferior surfaces of medial cuneiform and adjacent surfaces on base of first metatarsal.

Nerve supply
Deep fibular nerve L4, 5.

Blood supply
Anterior tibial artery
(from popliteal artery, a continuation of femoral artery).

Action
Dorsiflexes foot at ankle joint. Inverts foot. Dynamic support of medial arch of foot.

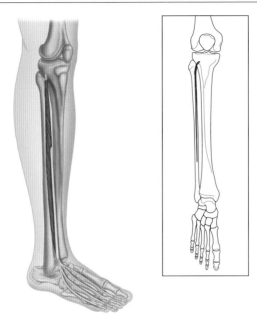

Extensor Digitorum Longus

Latin, *extendere*, to extend; *digitorum*, of the toes; *longus*, long.

Origin
Proximal one-half of medial surface of fibula and related surface of lateral tibial condyle.

Insertion
Along dorsal surface of the four lateral toes. Each tendon divides, to attach to bases of middle and distal phalanges.

Nerve supply
Deep fibular nerve L5, S1.

Blood supply
Anterior tibial artery (from popliteal artery, a continuation of femoral artery).

Action
Extends lateral four toes and dorsiflexes foot.

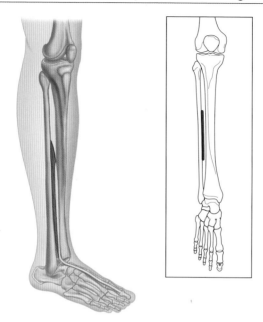

Extensor Hallucis Longus

Latin, *extendere*, to extend; *hallucis*, of the great toe; *longus*, long.

Origin
Middle one-half of medial surface of fibula and adjacent interosseous membrane.

Insertion
Base of distal phalanx of great toe.

Nerve supply
Deep fibular nerve L5, S1.

Blood supply
Anterior tibial artery
(from popliteal artery, a continuation of femoral artery).

Action
Extends great toe. Dorsiflexes foot.

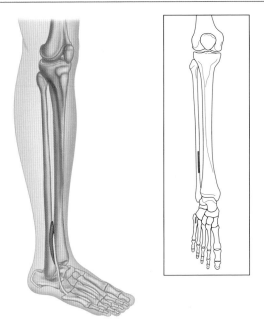

Fibularis Tertius

Latin, *fibula*, pin/buckle; *tertius*, third.

Origin
Distal part of medial surface of fibula.

Insertion
Dorsomedial surface of base of fifth metatarsal.

Nerve supply
Deep fibular nerve L5, S1.

Blood supply
Anterior tibial artery
(from popliteal artery, a continuation of femoral artery).

Action
Dorsiflexes and everts foot.

Muscles of the Posterior Compartment of the Leg—Superficial Layer

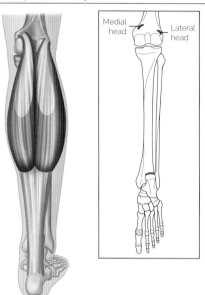

Gastrocnemius

Greek, *gaster*, stomach; *kneme*, lower leg.

Origin
Medial head
Posterior surface of distal femur just superior to medial condyle.
Lateral head
Upper posterolateral surface of lateral femoral condyle.

Insertion
Posterior surface of calcaneus via the Achilles tendon.

Nerve supply
Tibial nerve S1, 2.

Blood supply
Sural branches of popliteal artery
(a continuation of femoral artery).
Posterior tibial artery
(from popliteal artery).

Action
Plantar flexes foot. Flexes knee. It is a main propelling force in walking and running.

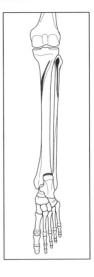

Soleus

Latin, *solea*, leather sole/sandal/sole (fish).

Origin
Posterior aspect of fibular head and adjacent surfaces of neck and proximal shaft. Soleal line and medial border of tibia. Tendinous arch between tibial and fibular attachments.

Insertion
Posterior surface of calcaneus via the Achilles tendon.

Nerve supply
Tibial nerve S1, 2.

Blood supply
Posterior tibial artery (from popliteal artery).
Sural branches of popliteal artery and fibular artery via posterior tibial artery.

Action
Plantar flexes foot. Soleus is frequently in contraction during standing, to prevent the body falling forward at the ankle joint. Thus, it helps to maintain an upright posture.

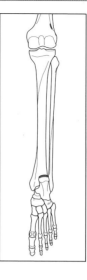

Plantaris

Latin, *plantaris*, relating to the sole.

Tibial nerve S1, 2.

Origin
Lower part of lateral supracondylar line of femur and oblique popliteal ligament of knee joint.

Insertion
Posterior surface of calcaneus via the Achilles tendon.

Blood supply
Sural branches of popliteal artery
(a continuation of femoral artery).

Action
Plantar flexes foot. Flexes knee.

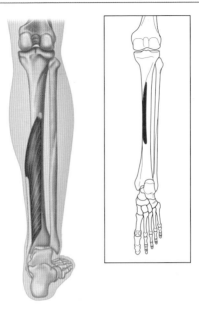

Flexor Digitorum Longus

Latin, *flectere*, to bend; *digitorum*, of the toes; *longus*, long.

Origin
Medial side of posterior surface of tibia, below soleal line.

Insertion
Plantar surfaces of bases of distal phalanges of lateral four toes.

Nerve supply
Tibial nerve S2, 3.

Blood supply
Posterior tibial artery
(from popliteal artery).

Action
Flexes lateral four toes (enabling the foot to firmly grip the ground when walking).

Muscles of the Posterior Compartment of the Leg—Intermediate Layer

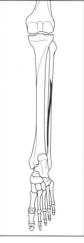

Flexor Hallucis Longus

Latin, *flectere*, to bend; *hallucis*, of the great toe; *longus*, long.

Origin
Lower two-thirds of posterior surface of fibula and adjacent interosseous membrane.

Insertion
Plantar surface of base of distal phalanx of great toe.

Nerve supply
Tibial nerve S2, 3.

Blood supply
Fibular artery
via posterior tibial artery (from popliteal artery).

Action
Flexes great toe, and is important in the final propulsive thrust of the foot during walking.

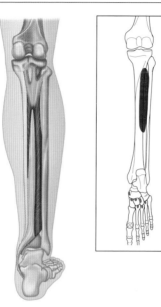

Tibialis Posterior

Latin, *tibialis*, relating to the shin; *posterior*, at the back.

Origin
Posterior surfaces of interosseous membrane and adjacent regions of tibia and fibula.

Insertion
Mainly to tuberosity of navicular and adjacent region of medial cuneiform.

Nerve supply
Tibial nerve L4, 5.

Blood supply
Fibular artery
via posterior tibial artery (from popliteal artery).

Action
Inverts and plantar flexes foot. Support of medial arch of the foot during walking.

Muscles of the Posterior Compartment of the Leg—Deep Layer

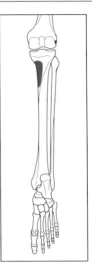

Popliteus

Latin, *poples*, knee, ham.

Origin
Lateral femoral condyle.

Insertion
Posterior surface of proximal tibia.

Nerve supply
Tibial nerve L4, 5, S1.

Blood supply
Sural and medial inferior genicular branches of popliteal artery
(a continuation of femoral artery).

Action
Stabilizes and unlocks the knee joint.

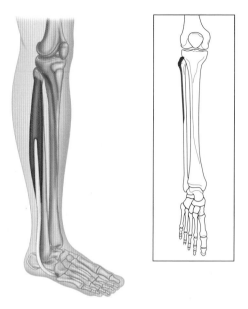

Fibularis Longus

Latin, *fibula*, pin/buckle; *longus*, long.

Origin
Upper two-thirds of lateral surface of fibula, head of fibula, and occasionally lateral tibial condyle.

Insertion
Lateral side of distal end of medial cuneiform. Base of first metatarsal.

Superficial fibular nerve L5, S1, 2.

Blood supply
Fibular artery
via posterior tibial artery (from popliteal artery).

Action
Everts and plantar flexes foot. Supports arches of foot.

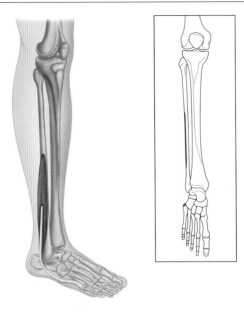

Fibularis Brevis

Latin, *fibula*, pin/buckle; *brevis*, short.

Origin
Lower two-thirds of lateral surface of shaft of fibula.

Insertion
Lateral tubercle at base of fifth metatarsal.

Nerve supply
Superficial fibular nerve L5, S1, 2.

Blood supply
Fibular artery
via posterior tibial artery (from popliteal artery).

Action
Everts foot.

Muscles of the Foot

The muscles acting on the foot can be divided into the extrinsic and intrinsic muscles. The extrinsic muscles arise from the anterior, lateral, and posterior compartments of the leg. They are mainly responsible for eversion, inversion, and plantar flexion of the foot.

The intrinsic muscles of the foot are mainly situated in the plantar region, or the sole of the foot. The sole can be described as consisting of an aponeurosis and then four muscle layers.

The plantar aponeurosis, also called the plantar fascia, is a fibrous flat sheet that lies deep to the superficial fascia of the sole and covers the first layer of muscles. It is attached to the calcaneus posteriorly and sends slips to each toe.

The muscular layers of the sole are:

- First layer, consisting of **abductor hallucis**, **flexor digitorum brevis**, and **abductor digiti minimi**.
- Second layer, consisting of **quadratus plantae** and **lumbricals**.
- Third layer, consisting of **flexor hallucis brevis**, **adductor hallucis**, and **flexor digiti minimi brevis**.
- Fourth layer, consisting of **dorsal** and **plantar interossei**.

Like the hand, the foot has lumbrical and interosseous muscles, but their functions are far less important. The **lumbricals** arise from tendons of flexor digitorum longus in the sole of the foot, and the **interossei** from the metatarsal bones. Their delicate tendons insert into the extensor expansions of the second to fifth toes, and their action is to flex the metatarsophalangeal joints and to weakly extend the interphalangeal joints.

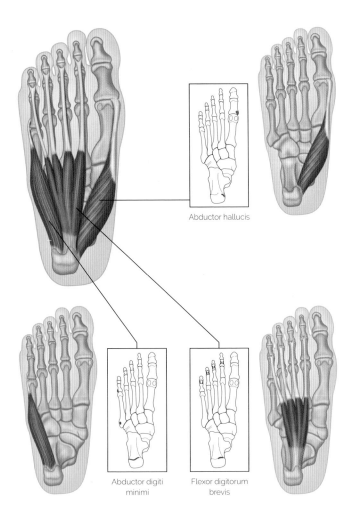

Abductor hallucis

Abductor digiti minimi

Flexor digitorum brevis

Abductor Hallucis

Latin, *abducere*, to lead away from; *hallucis*, of the great toe.

Origin
Medial process of calcaneal tuberosity.

Insertion
Medial side of base of proximal phalanx of great toe.

Nerve supply
Medial plantar nerve from tibial nerve S1–3.

Blood supply
Medial plantar artery
(from posterior tibial artery).

Action
Abducts and flexes great toe at metatarsophalangeal joint.

Flexor Digitorum Brevis

Latin, *flectere*, to bend; *digitorum*, of the toes; *brevis*, short.

Origin
Medial process of calcaneal tuberosity and plantar aponeurosis.

Insertion
Sides of plantar surfaces of middle phalanges of lateral four toes.

Nerve supply
Medial plantar nerve from tibial nerve S1–3.

Blood supply
Medial plantar artery
(from posterior tibial artery).

Action
Flexes lateral four toes at proximal interphalangeal joint.

Abductor Digiti Minimi

Latin, *abducere*, to lead away from; *digiti*, of the toe; *minimi*, of the smallest.

Origin
Lateral and medial processes of calcaneal tuberosity, and band of connective tissue connecting calcaneus with base of fifth metatarsal.

Insertion
Lateral side of base of proximal phalanx of little toe.

Nerve supply
Lateral plantar nerve from tibial nerve S1–3.

Blood supply
Lateral plantar artery
(from posterior tibial artery).

Action
Abducts fifth toe at metatarsophalangeal joint.

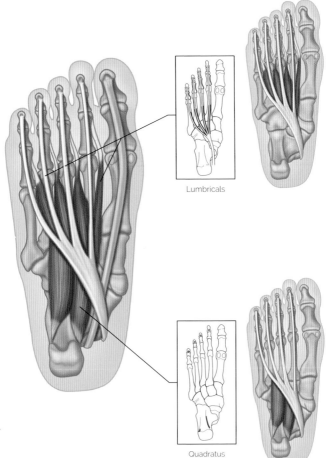

Lumbricals

Quadratus plantae

Quadratus Plantae

Latin, *quadratus*, squared; *plantae*, of the sole.

Origin
Medial surface of calcaneus and lateral process of calcaneal tuberosity.

Insertion
Lateral border of tendon of flexor digitorum longus in proximal sole of foot.

Nerve supply
Lateral plantar nerve from tibial nerve S1–3.

Blood supply
Lateral plantar artery
(from posterior tibial artery).

Action
Flexes distal phalanges of second to fifth toes. Modifies oblique line of pull of flexor digitorum longus tendons, to bring it in line with long axis of foot.

Lumbricals

Latin, *lumbricus*, earthworm.

Origin
First lumbrical
Medial side of tendon of flexor digitorum longus associated with second toe.
Second to fourth lumbricals
Adjacent tendons of flexor digitorum longus.

Insertion
Medial free margins of extensor hoods of second to fifth toes.

Nerve supply
First lumbrical
Medial plantar nerve from tibial nerve.
Lateral three lumbricals
Lateral plantar nerve from tibial nerve S2, 3.

Blood supply
First lumbrical
Medial plantar artery
Second to fourth lumbricals
Lateral plantar artery
(from posterior tibial artery).

Action
Flex metatarsophalangeal joint and extend interphalangeal joints.

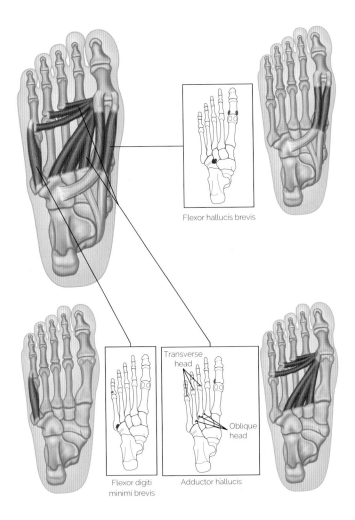

Flexor hallucis brevis

Flexor digiti minimi brevis

Transverse head

Oblique head

Adductor hallucis

Flexor Hallucis Brevis

Latin, *flectere*, to bend; *hallucis*, of the great toe; *brevis*, short.

Origin
Medial part of plantar surface of cuboid, and adjacent part of lateral cuneiform. Tendon of tibialis posterior.

Insertion
Lateral and medial sides of base of proximal phalanx of great toe.

Nerve supply
Medial plantar nerve from tibial nerve S1, 2.

Blood supply
Medial plantar artery
(from posterior tibial artery).

Action
Flexes metatarsophalangeal joint of great toe.

Adductor Hallucis

Latin, *adducere*, to lead to; *hallucis*, of the great toe.

Origin
Transverse head
Ligaments associated with metatarsophalangeal joints of lateral three toes.
Oblique head
Bases of second to fourth metatarsals; sheath covering fibularis longus tendon.

Insertion
Lateral side of base of proximal phalanx of great toe.

Nerve supply
Lateral plantar nerve from tibial nerve S2, 3.

Blood supply
Medial plantar artery
(from posterior tibial artery).

Action
Adducts great toe at metatarsophalangeal joint.

Flexor Digiti Minimi Brevis

Latin, *flectere*, to bend; *digiti*, of the toe; *minimi*, of the smallest; *brevis*, short.

Origin
Base of fifth metatarsal and sheath of fibularis longus tendon.

Insertion
Lateral side of base of proximal phalanx of little toe.

Nerve supply
Lateral plantar nerve from tibial nerve S2, 3.

Blood supply
Lateral plantar artery
(from posterior tibial artery).

Action
Flexes little toe at metatarsophalangeal joint.

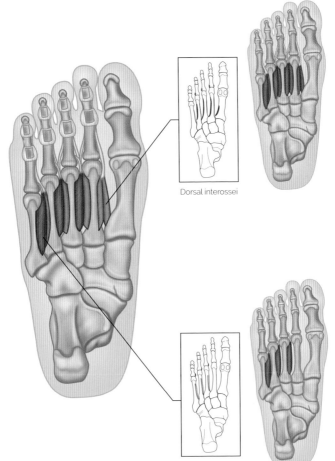

Dorsal interossei

Plantar interossei

Dorsal Interossei

Latin, *dorsalis*, relating to the back; *interosseus*, between bones.

Origin
Sides of adjacent metatarsals.

Insertion
Extensor hoods and bases of proximal phalanges of second to fourth toes.

Nerve supply
Lateral plantar nerve from tibial nerve; first and second dorsal interossei also innervated by deep fibular nerve S2, 3.

Blood supply
Dorsal metatarsal arteries via arcuate artery of foot (from dorsalis pedis artery, a continuation of anterior tibial artery).

Action
Abduct second to fourth toes at metatarsophalangeal joints. Resist extension of metatarsophalangeal joints and flexion of interphalangeal joints.

Plantar Interossei

Latin, *plantaris*, relating to the sole; *interosseus*, between bones.

Origin
Bases and medial sides of third to fifth metatarsals.

Insertion
Extensor hoods and bases of proximal phalanges of third to fifth toes.

Nerve supply
Lateral plantar nerve from tibial nerve S2, 3.

Blood supply
Plantar metatarsal arteries via plantar arch of lateral plantar artery (from posterior tibial artery).

Action
Adduct third to fifth toes at metatarsophalangeal joints. Resist extension of metatarsophalangeal joints and flexion of interphalangeal joints.

Muscles of the Dorsal Aspect of the Foot

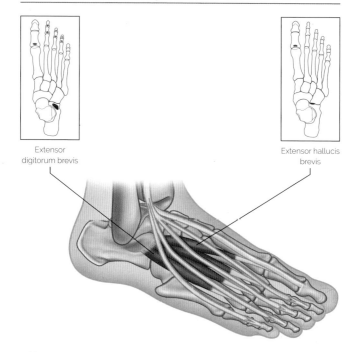

Extensor digitorum brevis

Extensor hallucis brevis

Extensor Digitorum Brevis and Extensor Hallucis Brevis

Latin, *extendere*, to extend; *digitorum*, of the toes; *hallucis*, of the great toe; *brevis*, short.

Origin
Superolateral surface of calcaneus.

Insertion
Extensor digitorum brevis
Lateral sides of tendons of extensor digitorum longus of second to fourth toes.
Extensor hallucis brevis
Base of proximal phalanx of great toe.

Deep fibular nerve S1, 2.

Blood supply
Dorsal pedis artery
(a continuation of anterior tibial artery).

Action
Extensor digitorum brevis
Extends second to fourth toes.
Extensor hallucis brevis
Extends metatarsophalangeal joint of great toe.

Dermatomes and Sensory Nerve Supply

Sensation from the skin is transferred to the spinal cord and hence to the brain by afferent nerve fibers (page 32) which are part of the mixed, motor, and sensory nerves, which make up the somatic peripheral nervous system.

All somatic nerves arise from one or more spinal segments and supply specific areas of skin. An area supplied by single spinal segment is called a *dermatome* but the nerves supplying a single dermatome may be carried in one or more individual nerves.

A good example of this is the C5 dermatome in the upper limb, which is supplied by the C5 fibers carried in both the upper lateral cutaneous nerve of the arm (axillary nerve) and the C5 fibers in the lower lateral cutaneous nerve of arm (radial nerve).

The dermatomes and the distribution of the individual nerves images that follow have been compiled by the publisher with the guidance and assistance of Dr Robert Whitaker, MA MD MChir FRCS FMAA, Anatomist, University of Cambridge.

Cutaneous Nerves of the Arm

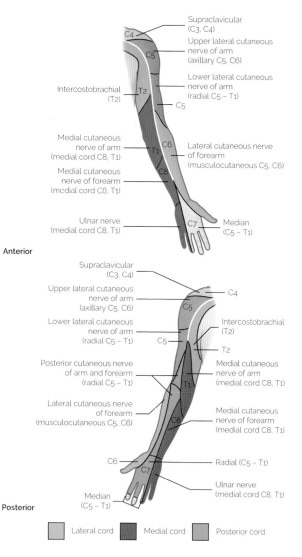

Supraclavicular
(C3, C4)

Upper lateral cutaneous
nerve of arm
(axillary C5, C6)

Lower lateral cutaneous
nerve of arm
(radial C5 – T1)

Intercostobrachial
(T2)

Medial cutaneous
nerve of arm
(medial cord C8, T1)

Lateral cutaneous nerve
of forearm
(musculocutaneous C5, C6)

Medial cutaneous
nerve of forearm
(medial cord C8, T1)

Ulnar nerve
(medial cord C8, T1)

Median
(C5 – T1)

Anterior

Supraclavicular
(C3, C4)

Upper lateral cutaneous
nerve of arm
(axillary C5, C6)

Lower lateral cutaneous
nerve of arm
(radial C5 – T1)

Posterior cutaneous nerve
of arm and forearm
(radial C5 – T1)

Intercostobrachial
(T2)

Medial cutaneous
nerve of arm
(medial cord C8, T1)

Lateral cutaneous nerve
of forearm
(musculocutaneous C5, C6)

Medial cutaneous
nerve of forearm
(medial cord C8, T1)

Radial (C5 – T1)

Ulnar nerve
(medial cord C8, T1)

Median
(C5 – T1)

Posterior

Lateral cord   Medial cord   Posterior cord

Cutaneous Nerves of the Leg

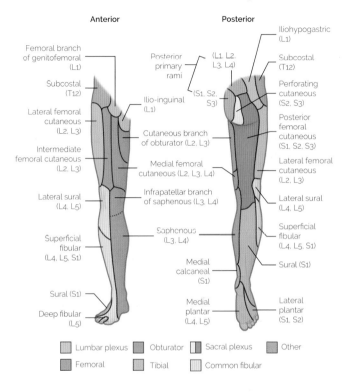

Anterior — Posterior

Femoral branch of genitofemoral (L1)

Subcostal (T12)

Lateral femoral cutaneous (L2, L3)

Intermediate femoral cutaneous (L2, L3)

Lateral sural (L4, L5)

Superficial fibular (L4, L5, S1)

Sural (S1)

Deep fibular (L5)

Posterior primary rami

(L1, L2, L3, L4)

(S1, S2, S3)

Ilio-inguinal (L1)

Cutaneous branch of obturator (L2, L3)

Medial femoral cutaneous (L2, L3, L4)

Infrapatellar branch of saphenous (L3, L4)

Saphenous (L3, L4)

Medial calcaneal (S1)

Medial plantar (L4, L5)

Iliohypogastric (L1)

Subcostal (T12)

Perforating cutaneous (S2, S3)

Posterior femoral cutaneous (S1, S2, S3)

Lateral femoral cutaneous (L2, L3)

Lateral sural (L4, L5)

Superficial fibular (L4, L5, S1)

Sural (S1)

Lateral plantar (S1, S2)

Lumbar plexus   Obturator   Sacral plexus   Other

Femoral   Tibial   Common fibular

Anterior and Posterior Dermatomes

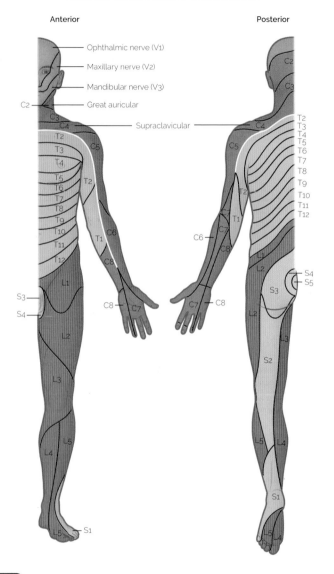

Anterior

- Ophthalmic nerve (V1)
- Maxillary nerve (V2)
- Mandibular nerve (V3)
- Great auricular
- Supraclavicular

C2
C3
C4
T2
T3
T4
T5
T6
T7
T8
T9
T10
T11
T12
C5
T2
C6
T1
C8
L1
S3
S4
L2
L3
L5
L4
C8
C7
L5
S1

Posterior

C2
C3
C4
C5
T2
T3
T4
T5
T6
T7
T8
T9
T10
T11
T12
T2
T1
C6
C7
C8
L1
L2
S4
S5
S3
C7
C8
L2
L3
S2
L5
L4
S1
L5
L4

Index of Muscles